Fundamentals of
Mechanics

Suresh Chandra
Mohit Kumar Sharma
Monika Sharma

FOUNDATION®
BOOKS

Bengaluru • Chennai • Delhi • Hyderabad • Kolkata • Mumbai
• Pune • Thiruvananthapuram

Published by
Cambridge University Press India Pvt. Ltd.
under the imprint of Foundation Books
Cambridge House, 4381/4 Ansari Road, Daryaganj, **New Delhi** 110 002

Cambridge University Press India Pvt. Ltd.
C-22, C-Block, Brigade M.M., K.R. Road, Jayanagar, **Bengaluru** 560 070
Plot No. 80, Service Industries, Shirvane, Sector-1, Nerul, **Navi Mumbai** 400 706
10 Raja Subodh Mullick Square, 2nd Floor, **Kolkata** 700 013
21/1 (New No. 49), 1st Floor, Model School Road, Thousand Lights, **Chennai** 600 006
House No. 3-5-874/6/4, (Near Apollo Hospital), Hyderguda, **Hyderabad** 500 029
Agarwal Pride, 'A' Wing, 1308 Kasba Peth, Near Surya Hospital, **Pune** 411 011
T.C. 25/ 2732 Lukes Lane, Ambujavilasam Road, **Thiruvananthapuram** 695 001

First Published 2014

ISBN 978-93-82993-40-7 (Paperback)

Published by Manas Saikia for Cambridge University Press India Pvt. Ltd. and printed at Akash Press, New Delhi 110 020

Contents

List of Figures

Preface

Fundamentals of Mechanics is the culmination of our years of experience in research and teaching the subject. It takes into account our close observations of student responses. Furthermore, it is a sincere attempt to make the study of mechanics enjoyable and arouse interest in the subject. Mechanics, as a subject of physics, is taught at BSc and BTech level. Professor Suresh Chandra conceptualized the book to fulfill requirements of science and engineering students of this level. The other two authors put in their extensive effort to give it a shape of a comprehensive textbook. There are chapters on vectors, laws of motion, conservation laws, inverse-square-law forces, harmonic oscillator, theory of relativity and non-inertial reference frames to match the title with the syllabuses of mechanics at different universities. The chapters are supplemented by ample number of exercises and diagrams which would help in better understanding of the subject. Multiple-choice questions and problems are provided at the end of each chapter to test progress after a lesson is taught.

It was only with the help, support and encouragement of a number of people that this project could come to fruition. First of all, we would like to express sincere thanks to our colleagues for their constructive criticism. Suresh Chandra is thankful to the authorities of the Lovely Professional University for their constant support. We also thank our family members for their kind cooperation and express our gratitude to Cambridge University Press India Pvt. Ltd. for publishing this book. Any suggestion for improvement of the book will be greatly appreciated and may be addressed through the publishers.

Suresh Chandra
Mohit Kumar Sharma
Monika Sharma

CHAPTER 1

VECTORS

The concept of vectors is very helpful in treatment of many physical phenomena in science and technology. A vector represents a physical quantity that has both direction as well as magnitude. A physical quantity which has only magnitude is known as a scalar.[1] Vector quantities follow some mathematical operations, such as addition, subtraction and multiplication. There is no direct operation for the division of one vector by another. Examples of vector quantities are displacement, velocity, acceleration, momentum, force, angular momentum, torque, electric field, magnetic field, electric dipole moment, and, so on. A vector quantity in this book will be denoted by putting an arrow above it.[2] For example, $\vec{r}$ denotes a vector quantity. Magnitude of this vector quantity will be denoted by r or by $|\vec{r}|$. A unit vector has a magnitude of 1. A unit vector parallel to $\vec{r}$ will be denoted by $\hat{r}$, which we read as 'r hat' or 'r caret'. Thus, we can write

$$\vec{r} = r\,\hat{r} = \hat{r}\,r$$

As mentioned earlier, a vector quantity has direction as well as magnitude. These features are independent of a coordinate system. The direction of a vector quantity can be defined, for example, with reference to a reference frame attached to a room, to the fixed stars, etc. For two vectors which are parallel to each other, one of them can be expressed in terms of the other. Suppose $\vec{A}$ and $\vec{B}$ are two parallel vectors with magnitudes 6 and 3 units, respectively, as shown in Figure 1.1(a). Then, we have $\vec{A} = 2\vec{B}$. When $\vec{A}$ and $\vec{B}$ are two vectors of equal magnitude, but anti parallel to each other, as shown in Figure 1.1(b). Then we have $\vec{A} = -\vec{B}$. In most of the discussion in this book, we shall consider Cartesian coordinates so that a vector $\vec{A}$ will be expressed as

$$\vec{A} = A_x\hat{i} + A_y\hat{j} + A_z\hat{k}$$

where A_x, A_y, A_z are, respectively, known as the x, y, z components of $\vec{A}$. Here, $\hat{i}, \hat{j}, \hat{k}$ denote the unit vectors along the positive directions of the x, y, z axes, respectively. Similarly, we have $\vec{B} = B_x\hat{i} + B_y\hat{j} + B_z\hat{k}$.

[1] There is no physical quantity which has only direction and no magnitude.

[2] Some authors write a vector quantity through a boldface, such as **r** instead of $\vec{r}$.

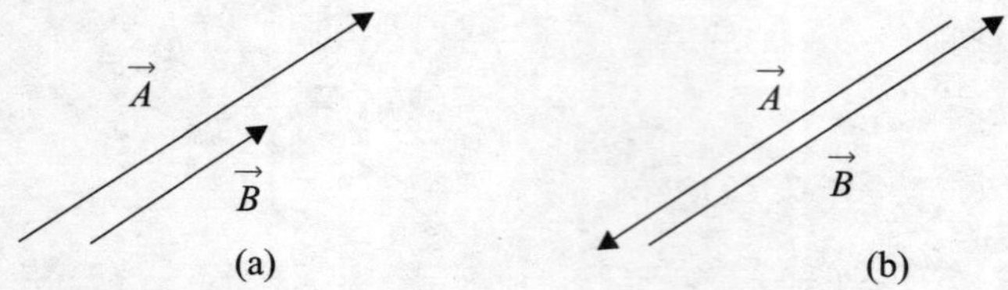

Figure 1.1: Graphical representation of vectors.

The magnitude of $\vec{A}$ is

$$A = |\vec{A}| = \sqrt{A_x^2 + A_y^2 + A_z^2}$$

1.1 Addition of vectors

Suppose $\vec{A}$ and $\vec{B}$ are two vectors. Their addition is expressed as

$$\vec{R} = \vec{A} + \vec{B}$$

When the vectors represent physical quantities, two vectors can be added only when they have the same dimensions. For example, a vector representing velocity cannot be added to another vector representing acceleration. (The same is true for subtraction of one vector from the other.) Addition of two vectors is commutative. That is

$$\vec{A} + \vec{B} = \vec{B} + \vec{A}$$

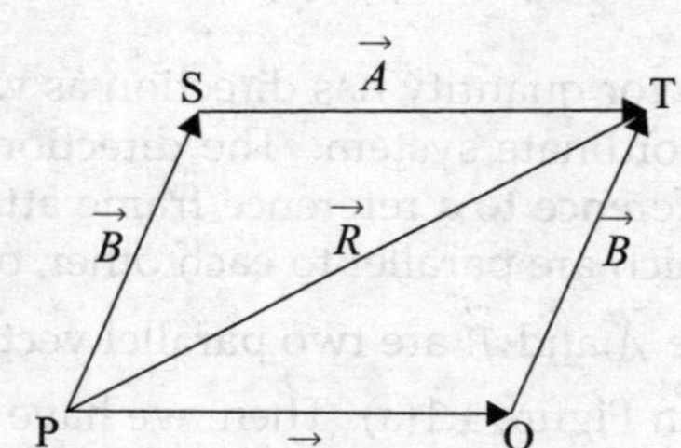

It can be proved in the following manner. In the given figure, we have

$$\vec{PQ} + \vec{QT} = \vec{PT} \qquad \text{or} \qquad \vec{A} + \vec{B} = \vec{R} \tag{1.1}$$

Further, we have

$$\vec{PS} + \vec{ST} = \vec{PT} \qquad \text{or} \qquad \vec{B} + \vec{A} = \vec{R} \tag{1.2}$$

From equations (1.1) and (1.2), we get

$$\vec{A} + \vec{B} = \vec{B} + \vec{A}$$

It shows that addition of two vectors is commutative. If we have

$$\vec{A} = A_x\hat{i} + A_y\hat{j} + A_z\hat{k} \quad \text{and} \quad \vec{B} = B_x\hat{i} + B_y\hat{j} + B_z\hat{k}$$

Then addition of them is

$$\vec{R} = \vec{A} + \vec{B} = (A_x\hat{i} + A_y\hat{j} + A_z\hat{k}) + (B_x\hat{i} + B_y\hat{j} + B_z\hat{k})$$

$$= (A_x + B_x)\hat{i} + (A_y + B_y)\hat{j} + (A_z + B_z)\hat{k}$$

Associative law

Addition of three vectors $\vec{A}$, $\vec{B}$ and $\vec{C}$ is associative. That is

$$\vec{A} + (\vec{B} + \vec{C}) = (\vec{A} + \vec{B}) + \vec{C}$$

Exercise 1.1: If $\vec{A} = A_x\hat{i} + A_y\hat{j} + A_z\hat{k}$, $\vec{B} = B_x\hat{i} + B_y\hat{j} + B_z\hat{k}$ and $\vec{C} = C_x\hat{i} + C_y\hat{j} + C_z\hat{k}$, show that $\vec{A} + (\vec{B} + \vec{C}) = (\vec{A} + \vec{B}) + \vec{C}$

Solution: We have

$$\vec{B} + \vec{C} = (B_x + C_x)\hat{i} + (B_y + C_y)\hat{j} + (B_z + C_z)\hat{k}$$

and therefore,

$$\vec{A} + (\vec{B} + \vec{C}) = (A_x + B_x + C_x)\hat{i} + (A_y + B_y + C_y)\hat{j} + (A_z + B_z + C_z)\hat{k} \tag{1.3}$$

Further, we have

$$\vec{A} + \vec{B} = (A_x + B_x)\hat{i} + (A_y + B_y)\hat{j} + (A_z + B_z)\hat{k}$$

and therefore,

$$(\vec{A} + \vec{B}) + \vec{C} = (A_x + B_x + C_x)\hat{i} + (A_y + B_y + C_y)\hat{j} + (A_z + B_z + C_z)\hat{k} \tag{1.4}$$

From equations (1.3) and (1.4), we have

$$\vec{A} + (\vec{B} + \vec{C}) = (\vec{A} + \vec{B}) + \vec{C}$$

It shows that addition of three vectors satisfies the associative law.

Distribution law

Addition of vectors satisfies the distribution law, expressed as

$$k(\vec{A} + \vec{B} + \vec{C}) = k\vec{A} + k\vec{B} + k\vec{C}$$

where k is a scalar quantity.

Exercise 1.2: If $\vec{A} = A_x\hat{i} + A_y\hat{j} + A_z\hat{k}$ and $\vec{B} = B_x\hat{i} + B_y\hat{j} + B_z\hat{k}$, show that $k(\vec{A} + \vec{B}) = k\vec{A} + k\vec{B}$.

Solution: We have

$$\vec{A} + \vec{B} = (A_x + B_x)\hat{i} + (A_y + B_y)\hat{j} + (A_z + B_z)\hat{k}$$

and

$$k(\vec{A} + \vec{B}) = (kA_x + kB_x)\hat{i} + (kA_y + kB_y)\hat{j} + (kA_z + kB_z)\hat{k} \tag{1.5}$$

Now,

$$k\vec{A} = kA_x\hat{i} + kA_y\hat{j} + kA_z\hat{k}$$

and

$$k\vec{B} = kB_x\hat{i} + kB_y\hat{j} + kB_z\hat{k}$$

Therefore,

$$k\vec{A} + k\vec{B} = (kA_x + kB_x)\hat{i} + (kA_y + kB_y)\hat{j} + (kA_z + kB_z)\hat{k} \tag{1.6}$$

From equations (1.5) and (1.6), we have

$$k(\vec{A} + \vec{B}) = k\vec{A} + k\vec{B}$$

1.1.1 Equilibrium under various forces

Suppose n forces $\vec{F}_1, \vec{F}_2, \ldots, \vec{F}_n$ are working simultaneously on a body. Here, the number of forces n is more than 1. When the resultant force working on the body is zero, that is,

$$\vec{F}_1 + \vec{F}_2 + \ldots + \vec{F}_n = 0$$

the situation is termed to be in equilibrium.

Exercise 1.3: Under the simultaneous action of four forces, a body is in equilibrium. When the three forces are $\vec{F}_1 = -2\hat{i} - 3\hat{j} + \hat{k}$, $\vec{F}_2 = 3\hat{i} + 2\hat{j} - 2\hat{k}$ and $\vec{F}_3 = -4\hat{i} - 5\hat{j} + 2\hat{k}$, find out the fourth force.

Solution: For the body to be in equilibrium, we have

$$\vec{F}_1 + \vec{F}_2 + \vec{F}_3 + \vec{F}_4 = 0 \quad \text{or} \quad \vec{F}_4 = -(\vec{F}_1 + \vec{F}_2 + \vec{F}_3)$$

Using the values of $\vec{F}_1$, $\vec{F}_2$, $\vec{F}_3$, we get

$$\vec{F}_4 = -[(-2\hat{\imath} - 3\hat{\jmath} + \hat{k}) + (3\hat{\imath} + 2\hat{\jmath} - 2\hat{k}) + (-4\hat{\imath} - 5\hat{\jmath} + 2\hat{k})]$$

$$= 3\hat{\imath} + 6\hat{\jmath} - \hat{k}$$

1.2 Subtraction of one vector from other

Suppose $\vec{A}$ and $\vec{B}$ are two vectors. The subtraction of $\vec{B}$ from $\vec{A}$ is expressed as

$$\vec{R} = \vec{A} - \vec{B}$$

For $\vec{A} = A_x\hat{\imath} + A_y\hat{\jmath} + A_z\hat{k}$ and $\vec{B} = B_x\hat{\imath} + B_y\hat{\jmath} + B_z\hat{k}$, we have

$$\vec{R} = \vec{A} - \vec{B} = (A_x\hat{\imath} + A_y\hat{\jmath} + A_z\hat{k}) - (B_x\hat{\imath} + B_y\hat{\jmath} + B_z\hat{k})$$

$$= (A_x - B_x)\hat{\imath} + (A_y - B_y)\hat{\jmath} + (A_z - B_z)\hat{k}$$

It is obvious that $(\vec{A} - \vec{B}) \neq (\vec{B} - \vec{A})$, showing that subtraction of one vector from the other is not commutative.

Exercise 1.4: For two vectors $\vec{A} = 2\hat{\imath} - 2\hat{\jmath} + 3\hat{k}$ and $\vec{B} = 3\hat{\imath} + 2\hat{\jmath} - \hat{k}$, calculate $2\vec{A} + 3\vec{B}$ and $2\vec{A} - \vec{B}$.

Solution: We have

$$2\vec{A} + 3\vec{B} = 2(2\hat{\imath} - 2\hat{\jmath} + 3\hat{k}) + 3(3\hat{\imath} + 2\hat{\jmath} - \hat{k}) = 13\hat{\imath} + 2\hat{\jmath} + 3\hat{k}$$

and

$$2\vec{A} - \vec{B} = 2(2\hat{\imath} - 2\hat{\jmath} + 3\hat{k}) - (3\hat{\imath} + 2\hat{\jmath} - \hat{k}) = \hat{\imath} - 6\hat{\jmath} + 7\hat{k}$$

1.3 Position vector

Let us consider a system of Cartesian coordinates as shown in Figure 1.2. Suppose the coordinates of a point P are (x, y, z). The position vector $\vec{r}$ of the point P relative to the origin O is

$$\vec{r} = x\hat{\imath} + y\hat{\jmath} + z\hat{k}$$

Here, $\hat{\imath}, \hat{\jmath}, \hat{k}$ denote the unit vectors along the x, y, z axes, respectively.

1.3.1 Displacement

Suppose a body moves from a point P with the position vector $\vec{r}_1 = x_1\hat{\imath} + y_1\hat{\jmath} + z_1\hat{k}$ to another point Q with position vector $\vec{r}_2 = x_2\hat{\imath} + y_2\hat{\jmath} + z_2\hat{k}$. The displacement of the body is

$$\vec{r} = \vec{r}_2 - \vec{r}_1 = (x_2\hat{\imath} + y_2\hat{\jmath} + z_2\hat{k}) - (x_1\hat{\imath} + y_1\hat{\jmath} + z_1\hat{k})$$

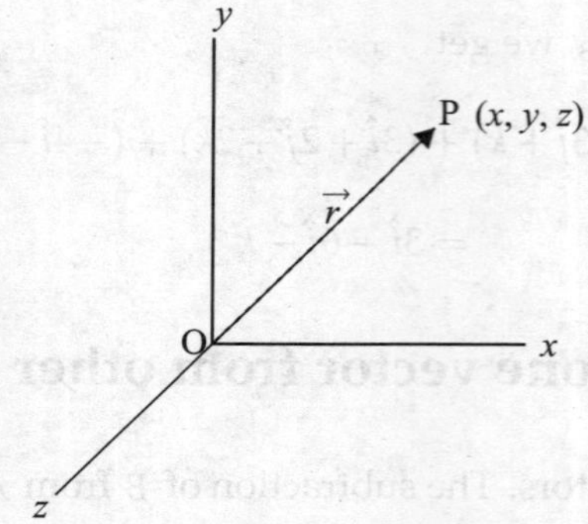

Figure 1.2: Cartesian coordinates of a point P.

$$= (x_2 - x_1)\hat{\imath} + (y_2 - y_1)\hat{\jmath} + (z_2 - z_1)\hat{k} = r_x\,\hat{\imath} + r_y\,\hat{\jmath} + r_z\,\hat{k}$$

where

$$r_x = x_2 - x_1 \qquad r_y = y_2 - y_1 \qquad r_z = z_2 - z_1$$

are known as the x, y, z components of the displacement.

Exercise 1.5: Suppose a body moves from a point P whose coordinates are (2, 4, 3) to another point Q whose coordinates are $(3, -2, 2)$. Calculate the displacement of the body.

Solution: With respect to the origin O of the coordinate system, we have

$$\vec{OP} = 2\hat{\imath} + 4\hat{\jmath} + 3\hat{k} \quad \text{and} \quad \vec{OQ} = 3\hat{\imath} - 2\hat{\jmath} + 2\hat{k}$$

Thus, the displacement of the body is

$$\vec{PQ} = \vec{OQ} - \vec{OP} = (3\hat{\imath} - 2\hat{\jmath} + 2\hat{k}) - (2\hat{\imath} + 4\hat{\jmath} + 3\hat{k})$$
$$= \hat{\imath} - 6\hat{\jmath} - \hat{k}$$

1.3.2 Momentum

Suppose a body (particle) of mass m is moving with velocity $\vec{v}$. The momentum $\vec{p}$ of the body is expressed as

$$\vec{p} = m\vec{v}$$

It is also known as linear momentum. The momentum is obviously along the direction of the velocity. If we have $\vec{v} = v_x\hat{\imath} + v_y\hat{\jmath} + v_z\hat{k}$ and $\vec{p} = p_x\hat{\imath} + p_y\hat{\jmath} + p_z\hat{k}$, then we have

$$p_x\hat{\imath} + p_y\hat{\jmath} + p_z\hat{k} = m(v_x\hat{\imath} + v_y\hat{\jmath} + v_z\hat{k})$$

and therefore, on comparing the coefficients of $\hat{\imath}, \hat{\jmath}, \hat{k}$ on two sides of this relation, we have

$$p_x = mv_x \qquad p_y = mv_y \qquad p_z = mv_z$$

Here, p_x, p_y, p_z are, respectively, known as the x, y, z components of momentum.

1.3.3 Electric force

Suppose a body (particle) having an electric charge q is placed in an external electric field $\vec{E}$. The force $\vec{F}$ acting on the charge is expressed as

$$\vec{F} = q\vec{E}$$

The force is obviously along the direction of the electric field when the charge is positive and in the opposite direction when the charge is negative.

1.4 Product of vectors

We have seen that the sum and difference of two vectors are vectors. But, in case of a product of two vectors, the resultant may be either a scalar quantity or a vector quantity. Accordingly, we have two types of multiplication of two vectors: (i) scalar product and (ii) vector product. In the case of a scalar product of two vectors, the resultant is a scalar quantity whereas in case of a vector product, the resultant is a vector quantity. For multiplication of two vectors, the vectors may have different dimensions.

1.4.1 Scalar product

Scalar product of two vectors $\vec{A}$ and $\vec{B}$ is expressed as (Figure 1.3)

$$\vec{A} \cdot \vec{B} = AB\cos\theta$$

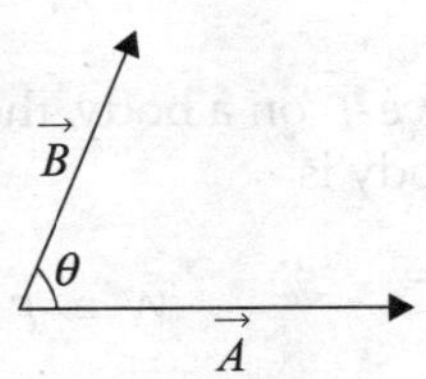

Figure 1.3: Two vectors inclined at an angle θ.

where A and B are magnitudes of the vectors $\vec{A}$ and $\vec{B}$, respectively, and θ denotes the angle between the two vectors. The $\vec{A} \cdot \vec{B}$ is read as $\vec{A}$ dot $\vec{B}$. As we use a dot sign between the two vectors, this product is also known as a dot product. The resultant of this product is a scalar quantity, and therefore it is known as a scalar product. As we have to take cosine of the angle between the vectors, it does not matter whether the angle θ is measured in the clockwise direction or in the anticlockwise direction.[3] Thus, we have

$$\vec{A} \cdot \vec{B} = \vec{B} \cdot \vec{A}$$

[3]Following the convention, the angle measured in the anticlockwise direction is taken as positive and in the clockwise direction as negative.

It shows that the scalar multiplication of two vectors is commutative. That is, the position of two vectors can be interchanged in the scalar product. Now, we have

$$\vec{A} \cdot \vec{A} = AA\cos 0^\circ = A^2$$

From the scalar product, for the unit vectors $\hat{\imath}, \hat{\jmath}, \hat{k}$ along the x, y, z axes, respectively, we get

$$\hat{\imath}.\hat{\imath} = 1 = \hat{\jmath}.\hat{\jmath} = \hat{k}.\hat{k} \qquad \text{and} \qquad \hat{\imath}.\hat{\jmath} = 0 = \hat{\jmath}.\hat{k} = \hat{k}.\hat{\imath}$$

For $\vec{A} = A_x\hat{\imath} + A_y\hat{\jmath} + A_z\hat{k}$ and $\vec{B} = B_x\hat{\imath} + B_y\hat{\jmath} + B_z\hat{k}$, we have

$$\vec{A} \cdot \vec{B} = A_xB_x + A_yB_y + A_zB_z$$

Exercise 1.6: We are given $\vec{A} = 2\hat{\imath} - 3\hat{\jmath} + \hat{k}$ and $\vec{B} = -\hat{\imath} - 2\hat{\jmath} + 4\hat{k}$. Calculate $\vec{A} \cdot \vec{B}$.

Solution: For the given $\vec{A}$ and $\vec{B}$, we have

$$\vec{A} \cdot \vec{B} = (2\hat{\imath} - 3\hat{\jmath} + \hat{k}).(-\hat{\imath} - 2\hat{\jmath} + 4\hat{k}) = -2 + 6 + 4 = 8$$

1.4.2 Some common examples of scalar product

Some common examples of scalar product in science and technology are as follows.

Work

Suppose on applying a force $\vec{F}$ on a body, the displacement of the body is $\vec{r}$. The work W done by the force on the body is

$$W = \vec{F} \cdot \vec{r}$$

When the displacement is along the direction of the applied force, we have $W = Fr$. Obviously, the work done is zero when the direction of displacement is perpendicular to the direction of the applied force. Example of such force is magnetic field.

Exercise 1.7: On application of a force $\vec{F} = 3\hat{\imath} + 2\hat{\jmath} + \hat{k}$, a body is displaced from a position (2, 3, 4) to (3, 2, 8). Calculate the amount of work done.

Solution: The displacement of the body is

$$\vec{r} = (3\hat{\imath} + 2\hat{\jmath} + 8\hat{k}) - (2\hat{\imath} + 3\hat{\jmath} + 4\hat{k}) = \hat{\imath} - \hat{\jmath} + 4\hat{k}$$

The work done is

$$W = \vec{F} \cdot \vec{r} = (3\hat{\imath} + 2\hat{\jmath} + \hat{k}).(\hat{\imath} - \hat{\jmath} + 4\hat{k}) = 5 \text{ unit}$$

Cosine law

For a triangle, the cosine law expresses cosine of an angle in terms of the lengths of the sides of the triangle. Suppose we want to find out the angle between the two vectors $\vec{P}$ and $\vec{Q}$ shown in the following figure:

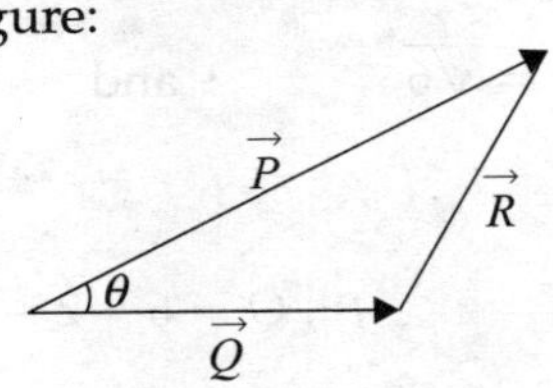

Let us write

$$\vec{R} = \vec{P} - \vec{Q}$$

Taking the scalar product, we have

$$\vec{R} \cdot \vec{R} = (\vec{P} - \vec{Q}).(\vec{P} - \vec{Q}) = \vec{P} \cdot \vec{P} - \vec{P} \cdot \vec{Q} - \vec{Q} \cdot \vec{P} + \vec{Q} \cdot \vec{Q}$$

Using the relation $\vec{P} \cdot \vec{Q} = \vec{Q} \cdot \vec{P}$, we have

$$R^2 = P^2 + Q^2 - 2\,\vec{P} \cdot \vec{Q} \qquad \text{or} \qquad R^2 = P^2 + Q^2 - 2PQ\cos\theta$$

where θ is the angle between the vectors $\vec{P}$ and $\vec{Q}$. Thus, we have

$$\cos\theta = \frac{P^2 + Q^2 - R^2}{2PQ}$$

This is known as the cosine law for a triangle.

The angle θ can also be obtained in the following manner. If $\vec{P} = P_x\hat{i} + P_y\hat{j} + P_z\hat{k}$ and $\vec{Q} = Q_x\hat{i} + Q_y\hat{j} + Q_z\hat{k}$, then their magnitudes are

$$P = \sqrt{P_x^2 + P_y^2 + P_z^2} \qquad \text{and} \qquad Q = \sqrt{Q_x^2 + Q_y^2 + Q_z^2}$$

and

$$\vec{P} \cdot \vec{Q} = P_xQ_x + P_yQ_y + P_zQ_z$$

From the relation

$$\vec{P} \cdot \vec{Q} = PQ\cos\theta \qquad \text{we have} \qquad \cos\theta = \frac{\vec{P} \cdot \vec{Q}}{PQ}$$

Using the values, we get

$$\cos\theta = \frac{P_xQ_x + P_yQ_y + P_zQ_z}{\sqrt{P_x^2 + P_y^2 + P_z^2}\,\sqrt{Q_x^2 + Q_y^2 + Q_z^2}}$$

Exercise 1.8: Obtain angle between two vectors expressed as $\vec{P} = 2\,\hat{\imath} - \hat{\jmath} + \hat{k}$ and $\vec{Q} = 3\,\hat{\imath} + 2\,\hat{\jmath} - 2\,\hat{k}$.

Solution: The magnitudes of vectors $\vec{P}$ and $\vec{Q}$, respectively, are

$$P = \sqrt{4+1+1} = \sqrt{6} \qquad \text{and} \qquad Q = \sqrt{9+4+4} = \sqrt{17}$$

Further, we have

$$\vec{P}\,.\,\vec{Q} = 6 - 2 - 2 = 2$$

If θ is the angle between the vectors $\vec{P}$ and $\vec{Q}$, we have

$$\cos\theta = \frac{\vec{P}\,.\,\vec{Q}}{PQ} = \frac{2}{\sqrt{6}\sqrt{17}} = 0.198$$

or

$$\theta = \cos^{-1} 0.198 = 78.6^\circ$$

1.4.3 Vector product

Vector product of two vectors $\vec{A}$ and $\vec{B}$ is expressed as

$$\vec{A} \times \vec{B} = AB\sin\theta\,\hat{n}$$

where A and B are magnitudes of the vectors $\vec{A}$ and $\vec{B}$, respectively, θ is the angle between the two vectors and $\hat{n}$ denotes the unit vector perpendicular to the plane containing both $\vec{A}$ and $\vec{B}$. The direction of $\hat{n}$ is decided with the help of the right hand. When the fingers of right hand are stretched from $\vec{A}$ towards $\vec{B}$, the direction of the thumb is the direction of $\hat{n}$. The $\vec{A} \times \vec{B}$ is read as $\vec{A}$ cross $\vec{B}$. As we use a cross sign between two vectors, the product is also known as cross product. The resultant of this product is a vector quantity and therefore, it is also known as a vector product. When we calculate $\vec{B} \times \vec{A}$, the direction of $\hat{n}$ is reversed. Thus, we have

$$\vec{A} \times \vec{B} = -\,\vec{B} \times \vec{A}$$

From the vector product, for unit vectors $\hat{\imath}$, $\hat{\jmath}$, $\hat{k}$ along the x, y, z axes, respectively, we get

$$\hat{\imath} \times \hat{\imath} = 0 = \hat{\jmath} \times \hat{\jmath} = \hat{k} \times \hat{k}$$

and

$$\hat{\imath} \times \hat{\jmath} = \hat{k} \qquad \hat{\jmath} \times \hat{k} = \hat{\imath} \qquad \hat{k} \times \hat{\imath} = \hat{\jmath}$$

For $\vec{A} = A_x\hat{\imath} + A_y\hat{\jmath} + A_z\hat{k}$ and $\vec{B} = B_x\hat{\imath} + B_y\hat{\jmath} + B_z\hat{k}$, we have

$$\vec{A} \times \vec{B} = (A_x\hat{\imath} + A_y\hat{\jmath} + A_z\hat{k}) \times (B_x\hat{\imath} + B_y\hat{\jmath} + B_z\hat{k})$$

$$= \hat{\imath}(A_yB_z - A_zB_y) - \hat{\jmath}(A_xB_z - A_zB_x) + \hat{k}(A_xB_y - A_yB_x)$$

It can be expressed in the matrix form as

$$\vec{A} \times \vec{B} = \begin{vmatrix} \hat{\imath} & \hat{\jmath} & \hat{k} \\ A_x & A_y & A_z \\ B_x & B_y & B_z \end{vmatrix}$$

Now, we have

$$\vec{A} \times \vec{A} = AA \sin 0^\circ \, \hat{n} = 0$$

Exercise 1.9: We are given $\vec{A} = \hat{\imath} - 2\,\hat{\jmath} + 3\,\hat{k}$ and $\vec{B} = 2\,\hat{\imath} + \hat{\jmath} - \hat{k}$. Calculate $\vec{A} \times \vec{B}$.

Solution: For the given $\vec{A}$ and $\vec{B}$, we have

$$\vec{A} \times \vec{B} = \begin{vmatrix} \hat{\imath} & \hat{\jmath} & \hat{k} \\ 1 & -2 & 3 \\ 2 & 1 & -1 \end{vmatrix} = -\hat{\imath} + 7\,\hat{\jmath} + 5\,\hat{k}$$

1.4.4 Some common examples of vector product

Some common examples of vector product in science and technology are as follows.

Area of a rectangle

Let us consider a rectangle PQRS whose adjacent sides PQ and PS can be expressed by the two vectors $\vec{A}$ and $\vec{B}$, as shown in the following figure.

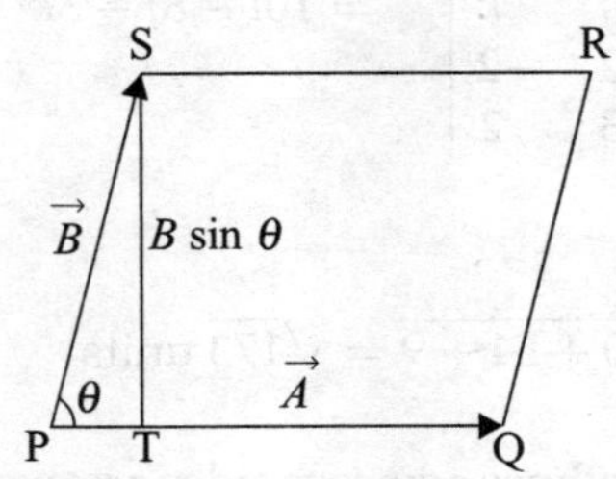

Draw a perpendicular ST on the line PQ from the point S. The area of the rectangle is

$$\text{Area} = \text{PQ} \times \text{ST} = A(B\sin\theta) = |\,\vec{A} \times \vec{B}\,|$$

Suppose

$$\vec{A} = A_x\,\hat{\imath} + A_y\,\hat{\jmath} + A_z\,\hat{k} \qquad \text{and} \qquad \vec{B} = B_x\,\hat{\imath} + B_y\,\hat{\jmath} + B_z\,\hat{k}$$

are two vectors. We have

$$\vec{A} \times \vec{B} = \begin{vmatrix} \hat{\imath} & \hat{\jmath} & \hat{k} \\ A_x & A_y & A_z \\ B_x & B_y & B_z \end{vmatrix}$$

$$= \hat{i}(A_yB_z - A_zB_y) - \hat{j}(A_xB_z - A_zB_x) + \hat{k}(A_xB_y - A_yB_x)$$

Therefore, we have

$$|\vec{A} \times \vec{B}| = \sqrt{(A_yB_z - A_zB_y)^2 + (A_xB_z - A_zB_x)^2 + (A_xB_y - A_yB_x)^2}$$

Exercise 1.10: Find the area of a rectangle shown below where the coordinates of points are A (1, −1,1), B(2, 1, −1), D (4, 2, 3).

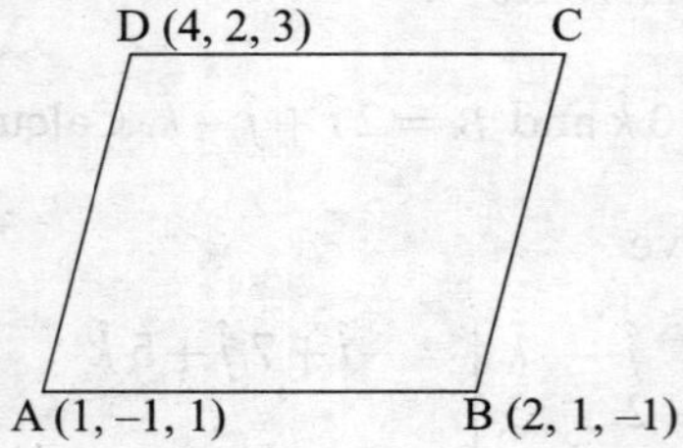

Solution: We have $\vec{OA} = \hat{i} - \hat{j} + \hat{k}$, $\vec{OB} = 2\,\hat{i} + \hat{j} - \hat{k}$, $\vec{OD} = 4\,\hat{i} + 2\,\hat{j} + 3\,\hat{k}$ and therefore

$$\vec{AB} = \vec{OB} - \vec{OA} = (2\hat{i} + \hat{j} - \hat{k}) - (\hat{i} - \hat{j} + \hat{k}) = \hat{i} + 2\hat{j} - 2\hat{k}$$

and

$$\vec{AD} = \vec{OD} - \vec{OA} = (4\hat{i} + 2\hat{j} + 3\hat{k}) - (\hat{i} - \hat{j} + \hat{k}) = 3\hat{i} + 3\hat{j} + 2\hat{k}$$

Thus, we have

$$\vec{AB} \times \vec{AD} = \begin{vmatrix} \hat{i} & \hat{j} & \hat{k} \\ 1 & 2 & -2 \\ 3 & 3 & 2 \end{vmatrix} = 10\hat{i} - 8\hat{j} - 3\hat{k}$$

Thus, area of the parallelepiped is

$$|\vec{AB} \times \vec{AD}| = \sqrt{100 + 64 + 9} = \sqrt{173} \text{ units}$$

Exercise 1.11: Find the area of a rectangle whose adjacent sides are expressed by two vectors:

$$\vec{A} = 4\hat{i} + 5\hat{j} - 3\hat{k} \qquad \text{and} \qquad \vec{B} = 2\hat{i} - 3\hat{j} + 4\hat{k}$$

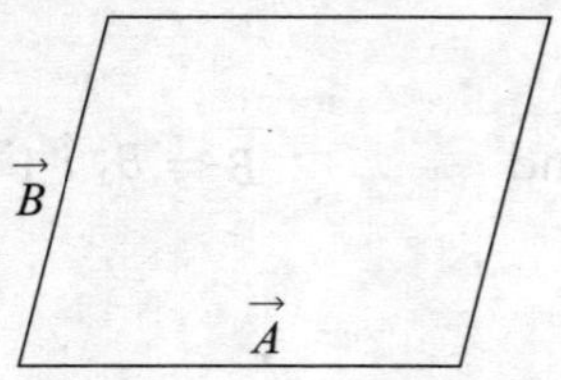

Solution: We have the vectors $\vec{A}$ and $\vec{B}$ as adjacent sides of a rectangle.

Thus, we have

$$\vec{A} \times \vec{B} = \begin{vmatrix} \hat{i} & \hat{j} & \hat{k} \\ 4 & 5 & -3 \\ 2 & -3 & 4 \end{vmatrix} = 11\hat{i} - 22\hat{j} - 22\hat{k}$$

Thus, the area of the rectangle is

$$|\vec{A} \times \vec{B}| = \sqrt{11^2 + 22^2 + 22^2} = 33 \text{ units}$$

Angular momentum

Suppose a body of mass m is moving with velocity $\vec{v}$, its momentum is $\vec{p} = m\vec{v}$. The body is at a point A whose position vector is $\vec{r}$, as shown in Figure 1.4, where both $\vec{p}$ and $\vec{r}$ lie in the plane of the paper. Let us consider an axis perpendicular to the plane of the paper and passing through a point O, referred to as the point of reference. Such axis is generally referred to as the axis of rotation. The angular momentum $\vec{J}$ of the body about O is expressed as

$$\vec{J} = \vec{r} \times \vec{p}$$

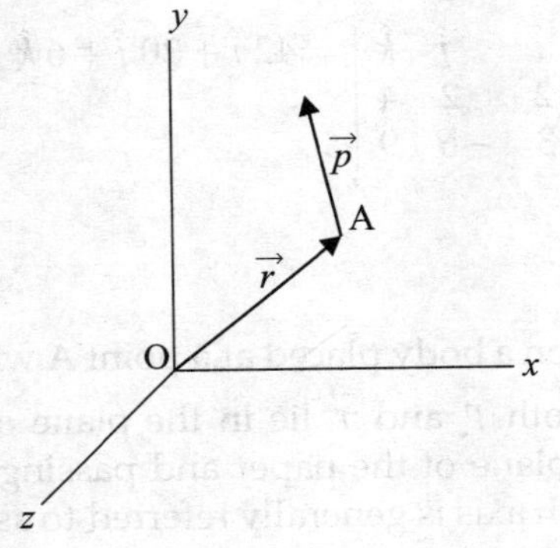

Figure 1.4: A body having momentum $\vec{p}$ is at a position $\vec{r}$.

The direction of $\vec{J}$ is perpendicular to the plane of the paper in the upward direction (in the sense of the vector product of two vectors).

Exercise 1.12: A body of mass 1.5 kg is moving with velocity $\vec{v} = 2\hat{i} - 4\hat{j} + 6\hat{k}$ when it is at a point P whose coordinates are (2, 3, 5). Find out the angular momentum of the body about an axis passing through a point Q whose coordinates are (4, 1, 1).

Solution: Here, the axis of rotation passes through the point Q. With respect to the origin O of the coordinate system, we have

$$\vec{OP} = 2\hat{i} + 3\hat{j} + 5\hat{k} \quad \text{and} \quad \vec{OQ} = 4\hat{i} + \hat{j} + \hat{k}$$

The position vector of P relative to Q is

$$\vec{r} = \vec{OP} - \vec{OQ} = (2\hat{i} + 3\hat{j} + 5\hat{k}) - (4\hat{i} + \hat{j} + \hat{k}) = -2\hat{i} + 2\hat{j} + 4\hat{k}$$

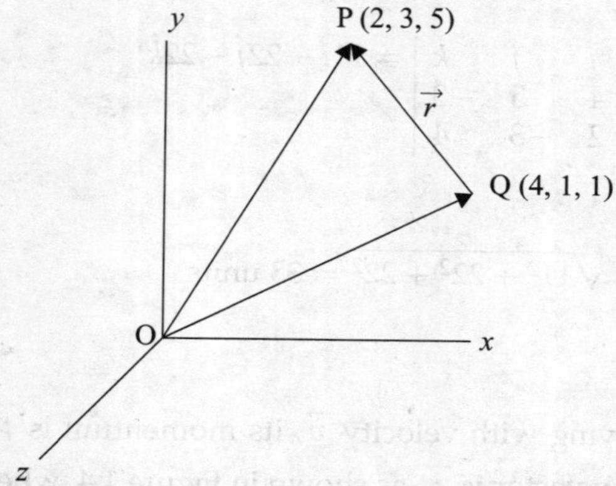

Momentum $\vec{p}$ of the body is

$$\vec{p} = m\vec{v} = 1.5(2\hat{\imath} - 4\hat{\jmath} + 6\hat{k}) = 3\hat{\imath} - 6\hat{\jmath} + 9\hat{k}$$

The angular momentum $\vec{J}$ is

$$\vec{J} = \vec{r} \times \vec{p} = \begin{vmatrix} \hat{\imath} & \hat{\jmath} & \hat{k} \\ -2 & 2 & 4 \\ 3 & -6 & 9 \end{vmatrix} = 42\,\hat{\imath} + 30\,\hat{\jmath} + 6\,\hat{k}$$

Torque

Suppose an external force $\vec{F}$ is applied on a body placed at a point A whose position vector is $\vec{r}$, as shown in Figure 1.5, where both $\vec{p}$ and $\vec{r}$ lie in the plane of the paper. Let us consider an axis perpendicular to the plane of the paper and passing through a point O, referred to as the point of reference. Such axis is generally referred to as the axis of rotation. The torque $\vec{\tau}$ of the force $\vec{F}$ about O is expressed as

$$\vec{\tau} = \vec{r} \times \vec{F}$$

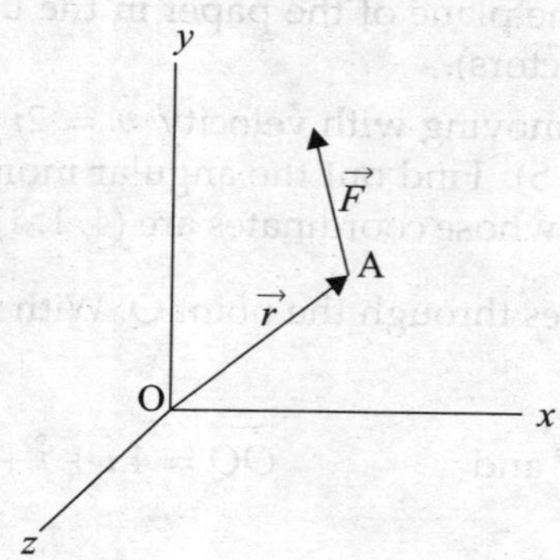

Figure 1.5: External force $\vec{F}$ is applied at a point A whose position vector is $\vec{r}$.

The direction of $\vec{\tau}$ is perpendicular to the plane of the paper in the upward direction (in the sense of the vector product of two vectors).

Exercise 1.13: A force $\vec{F} = 2\hat{\imath} - \hat{\jmath} + 4\hat{k}$ is acting at a point P whose coordinates are (5, −1, 2). Find out the torque of the force about an axis passing through a point Q whose coordinates are (2, 1, −3).

Solution: Here, the axis of rotation passes through the point Q. With respect to the origin O of the coordinate system, we have

$$\vec{OP} = 5\hat{\imath} - \hat{\jmath} + 2\hat{k} \qquad \text{and} \qquad \vec{OQ} = 2\hat{\imath} + \hat{\jmath} - 3\hat{k}$$

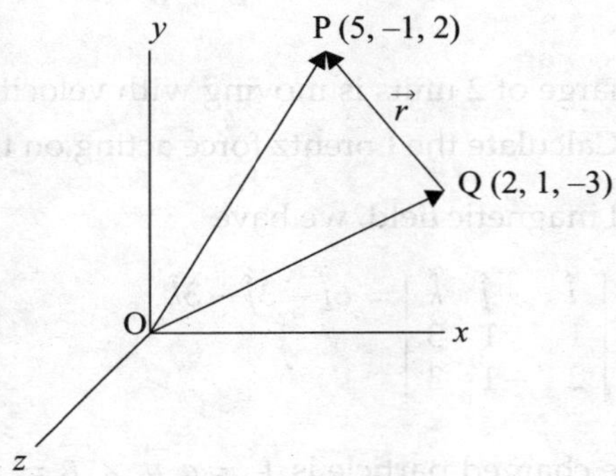

The position vector of P relative to Q is

$$\vec{r} = \vec{OP} - \vec{OQ} = (5\hat{\imath} - \hat{\jmath} + 2\hat{k}) - (2\hat{\imath} + \hat{\jmath} - 3\hat{k}) = 3\hat{\imath} - 2\hat{\jmath} + 5\hat{k}$$

The torque $\vec{\tau}$ of the force $\vec{F}$ is

$$\vec{\tau} = \vec{r} \times \vec{F} = \begin{vmatrix} \hat{\imath} & \hat{\jmath} & \hat{k} \\ 3 & -2 & 5 \\ 2 & -1 & 4 \end{vmatrix} = -3\hat{\imath} - 2\hat{\jmath} + \hat{k}$$

Lorentz force

Suppose a body having electric charge q is moving in a magnetic field $\vec{B}$ with velocity $\vec{v}$. The force $\vec{F}$, called the Lorentz force, acting on the body is

$$\vec{F} = q\,\vec{v} \times \vec{B}$$

The direction of force is perpendicular to the plane containing $\vec{v}$ and $\vec{B}$ both. Obviously, the direction of displacement is perpendicular to the direction of applied magnetic field. Consequently, the work done by the magnetic field is zero. It shows that a magnetic field changes the direction of motion of the charged particle, but does not change its energy.

When the direction of motion of the charged particle is either parallel or anti-parallel to the applied magnetic field, we have $\vec{v} \times \vec{B} = 0$ and the force acting on the charged particle is zero. Such a magnetic field is known as force-free magnetic field.

When both the electric field $\vec{E}$ and magnetic field $\vec{B}$ are present simultaneously, the force $\vec{F}$ acting on the charge q is

$$\vec{F} = q\,\vec{E} + q\,\vec{v} \times \vec{B}$$

It is quite possible that on the application of both electric field $\vec{E}$ and magnetic field $\vec{B}$, the net force acting on the charged particle is zero. That is

$$0 = q\,\vec{E} + q\,\vec{v} \times \vec{B} \qquad \text{or} \qquad \vec{E} = \vec{B} \times \vec{v}$$

Exercise 1.14: A particle carrying charge of 2 units is moving with velocity $\vec{v} = \hat{\imath} + \hat{\jmath} + 3\hat{k}$ in a magnetic field $\vec{B} = 2\hat{\imath} - \hat{\jmath} + 3\hat{k}$. Calculate the Lorentz force acting on the particle.

Solution: For the given velocity and magnetic field, we have

$$\vec{v} \times \vec{B} = \begin{vmatrix} \hat{\imath} & \hat{\jmath} & \hat{k} \\ 1 & 1 & 3 \\ 2 & -1 & 3 \end{vmatrix} = 6\hat{\imath} - 3\hat{\jmath} - 3\hat{k}$$

Thus, the Lorentz force acting on the charged particle is $\vec{F} = q\,\vec{v} \times \vec{B} = 2(6\hat{\imath} - 3\hat{\jmath} - 3\hat{k}) = 12\hat{\imath} - 12\hat{\jmath} - 6\hat{k}$.

Exercise 1.15: A charged particle is moving with velocity $\vec{v} = 2\hat{\imath} + \hat{\jmath} - \hat{k}$ in a magnetic field $\vec{B} = \hat{\imath} + \hat{\jmath} - 2\hat{k}$. Calculate the electric field applied so that the particle has no net force acting on it.

Solution: For the given velocity and magnetic field, we have

$$\vec{B} \times \vec{v} = \begin{vmatrix} \hat{\imath} & \hat{\jmath} & \hat{k} \\ 1 & 1 & -2 \\ 2 & 1 & -1 \end{vmatrix} = \hat{\imath} - 3\hat{\jmath} - \hat{k}$$

Thus, the electric field required is $\vec{E} = \hat{\imath} - 3\hat{\jmath} - \hat{k}$.

Sine law

For a triangle, the sine law relates an angle with the line opposite to it. Let us consider $\vec{C} = \vec{A} + \vec{B}$. All these three vectors $\vec{A}, \vec{B}, \vec{C}$ lie in a common plane. Now, we have

$$\vec{A} \times \vec{C} = \vec{A} \times (\vec{A} + \vec{B}) = \vec{A} \times \vec{A} + \vec{A} \times \vec{B}$$

Since $\vec{A} \times \vec{A} = 0$, we have

$$AC \sin(\vec{A}\,\vec{C})\,\hat{n} = AB \sin(\vec{A}\,\vec{B})\,\hat{n}$$

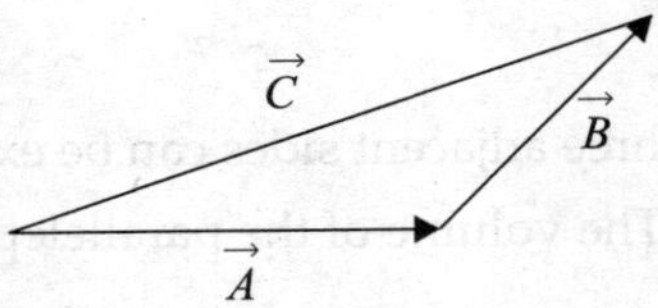

where, for example $(\vec{A}\,\vec{C})$ denotes the angle between the vectors $\vec{A}$ and $\vec{C}$. Note that in both the cross products, the unit vector $\hat{n}$ will be the same. Thus, we have

$$AC\sin(\vec{A}\,\vec{C}) = AB\sin(\vec{A}\,\vec{B})$$

and therefore, we get

$$\frac{\sin(\vec{A}\,\vec{C})}{B} = \frac{\sin(\vec{A}\,\vec{B})}{C}$$

This is known as the law of sine for a triangle.

1.4.5 Scalar triple product

Three vectors can be multiplied in two ways. One of them is scalar triple product whose resultant is a scalar quantity. Scalar triple product of three vectors $\vec{A}$, $\vec{B}$ and $\vec{C}$, represented by $[\vec{A}\,\vec{B}\,\vec{C}]$, is expressed as

$$\vec{A}\,.\,(\vec{B}\times\vec{C})$$

For $\vec{A} = A_x\hat{i} + A_y\hat{j} + A_z\hat{k}$, $\vec{B} = B_x\hat{i} + B_y\hat{j} + B_z\hat{k}$ and $\vec{C} = C_x\hat{i} + C_y\hat{j} + C_z\hat{k}$, $[\vec{A}\,\vec{B}\,\vec{C}]$ can be expressed by the following determinant:

$$[\vec{A}\,\vec{B}\,\vec{C}] = \begin{vmatrix} A_x & A_y & A_z \\ B_x & B_y & B_z \\ C_x & C_y & C_z \end{vmatrix}$$

$$= A_x(B_yC_z - B_zC_y) - A_y(B_xC_z - B_zC_x) + A_z(B_xC_y - B_yC_x)$$

This relation can be verified in the following manner. We have

$$\vec{B}\times\vec{C} = \hat{i}(B_yC_z - B_zC_y) - \hat{j}(B_xC_z - B_zC_x) + \hat{k}(B_xC_y - B_yC_x)$$

and therefore,

$$\vec{A}\,.\,(\vec{B}\times\vec{C}) = A_x(B_yC_z - B_zC_y) - A_y(B_xC_z - B_zC_x)$$

$$+A_z(B_xC_y - B_yC_x)$$

Volume of a parallelepiped

Let us consider a parallelepiped whose three adjacent sides can be expressed by three vectors $\vec{A}$, $\vec{B}$ and $\vec{C}$ as shown in Figure 1.6. The volume of the parallelepiped is

$$\text{Volume} = |\vec{A} \cdot (\vec{B} \times \vec{C})|$$

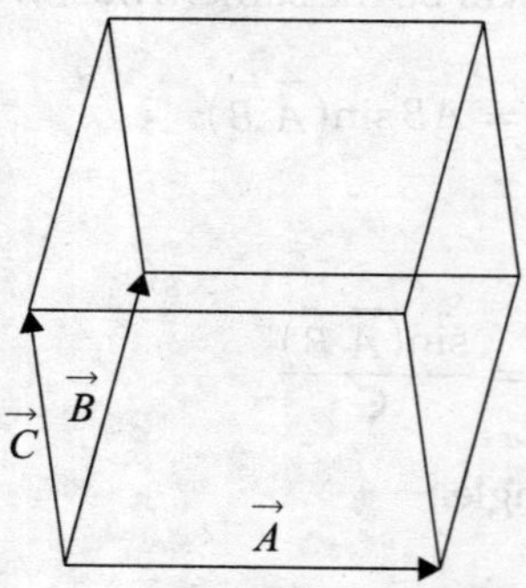

Figure 1.6: Three adjacent sides of a parallelepiped are represented by three vectors $\vec{A}$, $\vec{B}$ and $\vec{C}$.

For $\vec{A} = A_x\hat{i} + A_y\hat{j} + A_z\hat{k}$, $\vec{B} = B_x\hat{i} + B_y\hat{j} + B_z\hat{k}$ and $\vec{C} = C_x\hat{i} + C_y\hat{j} + C_z\hat{k}$, we have

$$\begin{aligned}\vec{A} \cdot (\vec{B} \times \vec{C}) &= \begin{vmatrix} A_x & A_y & A_z \\ B_x & B_y & B_z \\ C_x & C_y & C_z \end{vmatrix} \\ &= A_x(B_yC_z - B_zC_y) - A_y(B_xC_z - B_zC_x) + A_z(B_xC_y - B_yC_x)\end{aligned}$$

Since the volume is always positive, we have to take modulus of the vector product. It can also be written as

$$\text{Volume} = |\vec{A} \cdot (\vec{B} \times \vec{C})| = |\vec{B} \cdot (\vec{C} \times \vec{A})| = |\vec{C} \cdot (\vec{A} \times \vec{B})|$$

Exercise 1.16: Coordinates of points A, B, C, D in the following parallelepiped are (2, 3, 4), (1, 5, 0), (4, 2, 1) and (3, 4, 6), respectively. Calculate the volume of the parallelepiped.

Solution: We have $\vec{OA} = 2\hat{i} + 3\hat{j} + 4\hat{k}$, $\vec{OB} = \hat{i} + 5\hat{j}$, $\vec{OC} = 4\hat{i} + 2\hat{j} + \hat{k}$ and $\vec{OD} = 3\hat{i} + 4\hat{j} + 6\hat{k}$. The vectors for three adjacent sides are as follows:

$$\vec{AC} = \vec{OC} - \vec{OA} = (4\hat{i} + 2\hat{j} + \hat{k}) - (2\hat{i} + 3\hat{j} + 4\hat{k}) = 2\hat{i} - \hat{j} - 3\hat{k}$$

$$\vec{AB} = \vec{OB} - \vec{OA} = (\hat{i} + 5\hat{j}) - (2\hat{i} + 3\hat{j} + 4\hat{k}) = -\hat{i} + 2\hat{j} - 4\hat{k}$$

$$\vec{AD} = \vec{OD} - \vec{OA} = (3\hat{i} + 4\hat{j} + 6\hat{k}) - (2\hat{i} + 3\hat{j} + 4\hat{k}) = \hat{i} + \hat{j} + 2\hat{k}$$

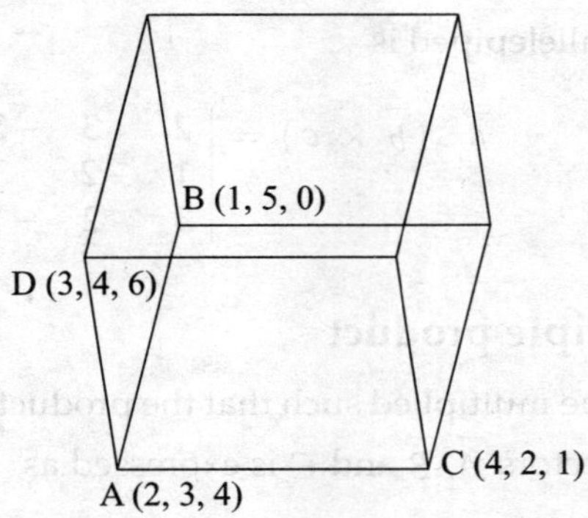

Volume of the parallelepiped is

$$\vec{AC}\,.\,(\vec{AB}\times\vec{AD}) = \begin{vmatrix} 2 & -1 & -3 \\ -1 & 2 & -4 \\ 1 & 1 & 2 \end{vmatrix} = 27 \text{ units}$$

Exercise 1.17: Three edges of a parallelepiped are expressed by three vectors: $\vec{a} = 2\hat{\imath} + 3\hat{\jmath} - 2\hat{k}$, $\vec{b} = \hat{\imath} - 2\hat{\jmath} + 3\hat{k}$ and $\vec{c} = 3\hat{\imath} + 2\hat{\jmath} - \hat{k}$. Find the area of the plane formed by $\vec{a}$ and $\vec{b}$, and the volume of the parallelepiped.

Solution: We have three edges as shown in the following figure.

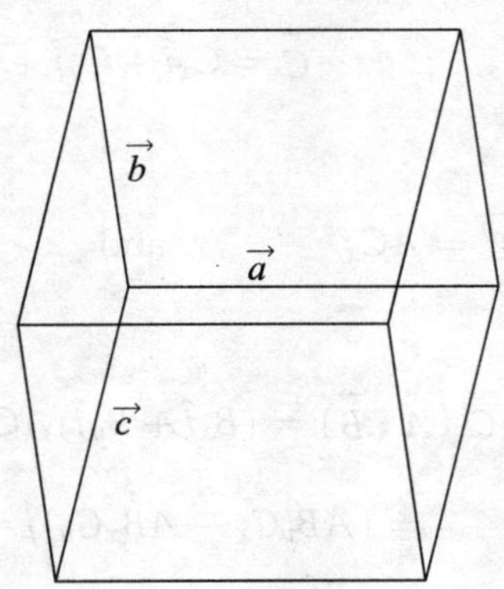

We have

$$\vec{a}\times\vec{b} = \begin{vmatrix} \hat{\imath} & \hat{\jmath} & \hat{k} \\ 2 & 3 & -2 \\ 1 & -2 & 3 \end{vmatrix} = 5\hat{\imath} - 8\hat{\jmath} - 7\hat{k}$$

Thus, the area formed by $\vec{a}$ and $\vec{b}$ is

$$|\,\vec{a}\times\vec{b}\,| = \sqrt{25+64+49} = \sqrt{138}$$

Volume of the parallelepiped is

$$\vec{a}\,.\,(\vec{b}\times\vec{c}) = \begin{vmatrix} 2 & 3 & -2 \\ 1 & -2 & 3 \\ 3 & 2 & -1 \end{vmatrix} = 6 \text{ units}$$

1.4.6 Vector triple product

Three vectors can be multiplied such that the product is a vector quantity. The vector triple product of three vectors $\vec{A}$, $\vec{B}$ and $\vec{C}$ is expressed as

$$\vec{A}\times(\vec{B}\times\vec{C}) = \vec{B}\,(\vec{A}\,.\,\vec{C}) - \vec{C}\,(\vec{A}\,.\,\vec{B})$$

This relation can be verified in the following manner. The Cartesian coordinate system can be aligned according to our choice. Suppose the x-axis is taken along the direction of the vector $\vec{A}$, so that

$$\vec{A} = A\,\hat{i}$$

After fixation of the x-axis, we can rotate the coordinate system about the x-axis in such a manner that the vector $\vec{B}$ is in the xy-plane, so that

$$\vec{B} = B_x\hat{i} + B_y\hat{j}$$

Now, the vector $\vec{C}$ is in the space and we have

$$\vec{C} = C_x\hat{i} + C_y\hat{j} + C_z\hat{k}$$

Now, we have

$$\vec{A}\,.\,\vec{C} = AC_x \qquad \text{and} \qquad \vec{A}\,.\,\vec{B} = AB_x$$

and therefore

$$\begin{aligned}\vec{B}\,(\vec{A}\,.\,\vec{C}) - \vec{C}\,(\vec{A}\,.\,\vec{B}) &= (B_x\hat{i} + B_y\hat{j})AC_x - (C_x\hat{i} + C_y\hat{j} + C_z\hat{k})AB_x \\ &= (AB_yC_x - AB_xC_y)\hat{j} - AB_xC_z\hat{k} \end{aligned} \tag{1.7}$$

Further, we have

$$\vec{B}\times\vec{C} = \begin{vmatrix} \hat{i} & \hat{j} & \hat{k} \\ B_x & B_y & 0 \\ C_x & C_y & C_z \end{vmatrix} = B_yC_z\hat{i} - B_xC_z\hat{j} + (B_xC_y - B_yC_x)\hat{k}$$

and

$$\vec{A}\times(\vec{B}\times\vec{C}) = \begin{vmatrix} \hat{i} & \hat{j} & \hat{k} \\ A & 0 & 0 \\ B_yC_z & -B_xC_z & B_xC_y - B_yC_x \end{vmatrix}$$

$$= (AB_yC_x - AB_xC_y)\hat{j} - AB_xC_z\hat{k} \tag{1.8}$$

From equations (1.7) and (1.8), we have

$$\vec{A} \times (\vec{B} \times \vec{C}) = \vec{B}\,(\vec{A} \cdot \vec{C}) - \vec{C}\,(\vec{A} \cdot \vec{B})$$

Exercise 1.18: If $\vec{A} = 2\hat{i} + 3\hat{j} + \hat{k}$, $\vec{B} = \hat{i} + \hat{j} + 2\hat{k}$ and $\vec{C} = 3\hat{i} - \hat{j} + 2\hat{k}$, prove that $\vec{A} \times (\vec{B} \times \vec{C}) = \vec{B}\,(\vec{A} \cdot \vec{C}) - \vec{C}\,(\vec{A} \cdot \vec{B})$.

Solution: We have

$$\vec{B} \times \vec{C} = \begin{vmatrix} \hat{i} & \hat{j} & \hat{k} \\ 1 & 1 & 2 \\ 3 & -1 & 2 \end{vmatrix} = 4\hat{i} + 4\hat{j} - 4\hat{k}$$

and

$$\vec{A} \times (\vec{B} \times \vec{C}) = \begin{vmatrix} \hat{i} & \hat{j} & \hat{k} \\ 2 & 3 & 1 \\ 4 & 4 & -4 \end{vmatrix} = -16\hat{i} + 12\hat{j} - 4\hat{k}$$

Now, we have

$$\vec{A} \cdot \vec{C} = (2\hat{i} + 3\hat{j} + \hat{k}).(3\hat{i} - \hat{j} + 2\hat{k}) = 5$$

and

$$\vec{A} \cdot \vec{B} = (2\hat{i} + 3\hat{j} + \hat{k}).(\hat{i} + \hat{j} + 2\hat{k}) = 7$$

Therefore,

$$\vec{B}\,(\vec{A} \cdot \vec{C}) - \vec{C}\,(\vec{A} \cdot \vec{B}) = 5(\hat{i} + \hat{j} + 2\hat{k}) - 7(3\hat{i} - \hat{j} + 2\hat{k})$$

$$= -16\hat{i} + 12\hat{j} - 4\hat{k}$$

It gives

$$\vec{A} \times (\vec{B} \times \vec{C}) = \vec{B}\,(\vec{A} \cdot \vec{C}) - \vec{C}\,(\vec{A} \cdot \vec{B})$$

1.5 Vector derivative

Suppose $\vec{A}\,(x)$ is a vector quantity which depends on a scalar variable x. At $x + \Delta x$, the vector is $\vec{A}\,(x + \Delta x)$. The change in the vector is

$$\Delta\vec{A} = \vec{A}\,(x + \Delta x) - \vec{A}\,(x)$$

Thus, we have

$$\frac{\Delta\vec{A}}{\Delta x} = \frac{\vec{A}\,(x + \Delta x) - \vec{A}\,(x)}{\Delta x}$$

Taking Δx tending to zero, we have

$$\lim_{\Delta x \to 0} \frac{\Delta \vec{A}}{\Delta x} = \frac{d\vec{A}}{dx}$$

Thus, we have

$$\frac{d\vec{A}}{dx} = \lim_{\Delta x \to 0} \frac{\vec{A}\,(x+\Delta x) - \vec{A}\,(x)}{\Delta x}$$

This is known as the derivative of $\vec{A}$ with respect to x. When $\vec{A}$ is expressed as

$$\vec{A} = A_x\hat{\imath} + A_y\hat{\jmath} + A_z\hat{k}$$

Then we have

$$\frac{d\vec{A}}{dt} = \frac{A_x}{dt}\,\hat{\imath} + \frac{A_y}{dt}\,\hat{\jmath} + \frac{A_z}{dt}\,\hat{k}$$

where we assume that the unit vectors do not change with respect to t. That is

$$\frac{d\hat{\imath}}{dt} = 0 \qquad \frac{d\hat{\jmath}}{dt} = 0 \qquad \frac{d\hat{k}}{dt} = 0$$

This is the case for Cartesian coordinates. However, in general (in other systems of coordinates) when we have $\vec{A} = A\hat{A}$, then

$$\frac{d\vec{A}}{dt} = \frac{dA}{dt}\,\hat{A} + A\,\frac{d\hat{A}}{dt}$$

That is, the unit vector $\hat{A}$ also changes.

1.5.1 Velocity

The velocity $\vec{v}$ of a body is defined as the rate of change of displacement $\vec{r}$ of the body. That is,

$$\vec{v} = \frac{d\vec{r}}{dt}$$

If we have $\vec{r} = x\hat{\imath} + y\hat{\jmath} + z\hat{k}$ and $\vec{v} = v_x\hat{\imath} + v_y\hat{\jmath} + v_z\hat{k}$, we get

$$v_x\hat{\imath} + v_y\hat{\jmath} + v_z\hat{k} = \frac{d}{dt}(x\hat{\imath} + y\hat{\jmath} + z\hat{k})$$

and therefore, on comparing the coefficients of $\hat{\imath}$, $\hat{\jmath}$, $\hat{k}$ on two sides of the relation, we have

$$v_x = \frac{dx}{dt} \qquad v_y = \frac{dy}{dt} \qquad v_z = \frac{dz}{dt}$$

Here, v_x, v_y, v_z are, respectively, known as the x, y, z components of the velocity.

Exercise 1.19: The displacement of a particle is expressed as $\vec{r} = (t^2 - 3t + 4)\hat{\imath} + (2t^3 - 4t)\hat{\jmath} + (2t+1)^2\hat{k}$, where t denotes the time. Calculate (i) the displacement and its magnitude at $t = 1$, (ii) velocity and its magnitude at $t = 1$.

Solution: The displacement of the particle is

$$\vec{r} = (t^2 - 3t + 4)\hat{\imath} + (2t^3 - 4t)\hat{\jmath} + (2t+1)^2\hat{k}$$

Thus, the displacement at $t = 1$ is

$$(\vec{r})_{t=1} = 2\hat{\imath} - 2\hat{\jmath} + 9\hat{k}$$

The magnitude of displacement of the particle at $t = 1$ is

$$(r)_{t=1} = \sqrt{4+4+81} = \sqrt{89} \text{ units}$$

The velocity of the particle is

$$\vec{v} = \frac{d\vec{r}}{dt} = (2t-3)\hat{\imath} + (6t^2 - 4)\hat{\jmath} + (8t+4)\hat{k}$$

Thus, the velocity at $t = 1$ is

$$(\vec{v})_{t=1} = -\hat{\imath} + 2\hat{\jmath} + 12\hat{k}$$

The magnitude of the velocity of the particle at $t = 1$ is

$$(v)_{t=1} = \sqrt{1+4+144} = \sqrt{149} \text{ units}$$

1.5.2 Acceleration

The acceleration $\vec{a}$ of a body is defined as the rate of change of velocity $\vec{v}$ of the body. That is

$$\vec{a} = \frac{d\vec{v}}{dt}$$

If we have $\vec{v} = v_x\hat{\imath} + v_y\hat{\jmath} + v_z\hat{k}$ and $\vec{a} = a_x\hat{\imath} + a_y\hat{\jmath} + a_z\hat{k}$, we have

$$a_x\hat{\imath} + a_y\hat{\jmath} + a_z\hat{k} = \frac{d}{dt}(v_x\hat{\imath} + v_y\hat{\jmath} + v_z\hat{k})$$

and therefore, on comparing the coefficients of $\hat{\imath}$, $\hat{\jmath}$, $\hat{k}$ on two sides of this relation, we have

$$a_x = \frac{dv_x}{dt} \qquad a_y = \frac{dv_y}{dt} \qquad a_z = \frac{dv_z}{dt}$$

Here, a_x, a_y, a_z are, respectively, known as the x, y, z components of the acceleration.

Exercise 1.20: The displacement of a body is expressed as

$$\vec{r} = \sin \omega t\, \hat{\imath} - \cos \omega t\, \hat{\jmath} + t^3\, \hat{k}$$

where t denotes the time and ω is a constant. Calculate (i) magnitude of displacement, (ii) velocity and its magnitude and (iii) acceleration and its magnitude.

Solution: The magnitude of displacement is

$$r = \sin^2 \omega t + \cos^2 \omega t + t^6 = 1 + t^6$$

The velocity $\vec{v}$ of the body is

$$\vec{v} = \frac{\mathrm{d}\vec{r}}{\mathrm{d}t} = \omega \cos \omega t\, \hat{\imath} + \omega \sin \omega t\, \hat{\jmath} + 3t^2\, \hat{k}$$

The magnitude of velocity is

$$v = \sqrt{\omega^2 \cos^2 \omega t + \omega^2 \sin^2 \omega t + 9t^4} = \sqrt{\omega^2 + 9t^4}$$

The acceleration $\vec{a}$ of the body is

$$\vec{a} = \omega^2 \sin \omega t\, \hat{\imath} - \omega^2 \cos \omega t\, \hat{\jmath} + 6t\, \hat{k}$$

The magnitude of acceleration is

$$a = \sqrt{\omega^4 \sin^2 \omega t + \omega^4 \cos^2 \omega t + 36t^2} = \sqrt{\omega^4 + 36t^2}$$

Exercise 1.21: The displacement of a body is expressed as

$$\vec{r} = t^2\hat{\imath} + (t+1)^2\hat{\jmath} + (t-1)^2\hat{k}$$

where t denotes the time. Calculate (i) magnitude of displacement, (ii) velocity and its magnitude and (iii) acceleration and its magnitude.

Solution: The magnitude of displacement is

$$r = |\,\vec{r}\,| = \sqrt{t^4 + (t+1)^4 + (t-1)^4}$$

$$= \sqrt{t^4 + (t^4 + 4t^3 + 6t^2 + 4t + 1) + (t^4 - 4t^3 + 6t^2 - 4t + 1)}$$

$$= \sqrt{3t^4 + 12t^2 + 2}$$

The velocity of the body is

$$\vec{v} = \frac{\mathrm{d}\vec{r}}{\mathrm{d}t} = 2t\hat{\imath} + 2(t+1)\hat{\jmath} + 2(t-1)\hat{k}$$

The magnitude of velocity is

$$v = \sqrt{4t^2 + 4(t+1)^2 + 4(t-1)^2} = \sqrt{12t^2 + 8}$$

The acceleration of the body is

$$\vec{a} = \frac{\mathrm{d}\vec{v}}{\mathrm{d}t} = 2\hat{\imath} + 2\hat{\jmath} + 2\hat{k}$$

The magnitude of acceleration is

$$a = \sqrt{4+4+4} = \sqrt{12} \text{ units}$$

1.5.3 Rate of change of unit vector $\hat{r}$

For motion of a body in the plane polar coordinates (r, θ), let us develop expressions for $(\mathrm{d}\hat{r}/\mathrm{d}t)$ and $(\mathrm{d}\hat{\theta}/\mathrm{d}t)$. The change in the direction of unit vector $\hat{r}$ is shown in Figure 1.7. The vectors $\hat{r}$ and $(\hat{r} + \Delta\hat{r})$ lie in the plane of the paper. We have

$$|\Delta\hat{r}| = |\hat{r}|\,\Delta\theta = \Delta\theta$$

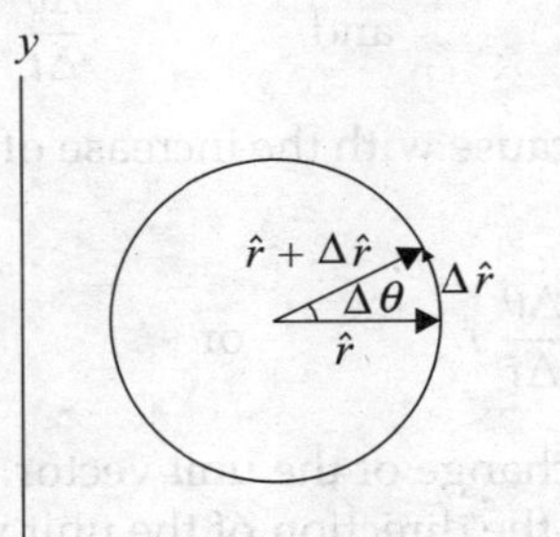

Figure 1.7: Change in the direction of unit vector $\hat{r}$.

as $|\hat{r}| = 1$. Thus, we have

$$\Delta\hat{r} = \Delta\theta\,\hat{\theta} \qquad \text{and} \qquad \frac{\Delta\hat{r}}{\Delta t} = \frac{\Delta\theta}{\Delta t}\,\hat{\theta}$$

The direction of the unit vector $\hat{\theta}$ is perpendicular to the plane of the paper in the upward direction. For Δt tending to zero, we have

$$\lim_{\Delta t \to 0} \frac{\Delta\hat{r}}{\Delta t} = \lim_{\Delta t \to 0} \frac{\Delta\theta}{\Delta t}\,\hat{\theta} \qquad \text{or} \qquad \frac{\mathrm{d}\hat{r}}{\mathrm{d}t} = \frac{\mathrm{d}\theta}{\mathrm{d}t}\,\hat{\theta} \tag{1.9}$$

Equation (1.9) gives the rate of change of the unit vector $\hat{r}$. Notice that the rate of change of the unit vector $\hat{r}$ is along the direction of the unit vector $\hat{\theta}$.

1.5.4 Rate of change of unit vector $\hat{\theta}$

In order to obtain the rate of change of unit vector $\hat{\theta}$, let us take the vectors $\hat{\theta}$ and $(\hat{\theta}+\Delta\hat{\theta})$ in the plane of the paper. The change in the direction of unit vector $\hat{\theta}$ is shown in Figure 1.8. We have

$$|\Delta\hat{\theta}| = |\hat{\theta}|\Delta\theta = \Delta\theta$$

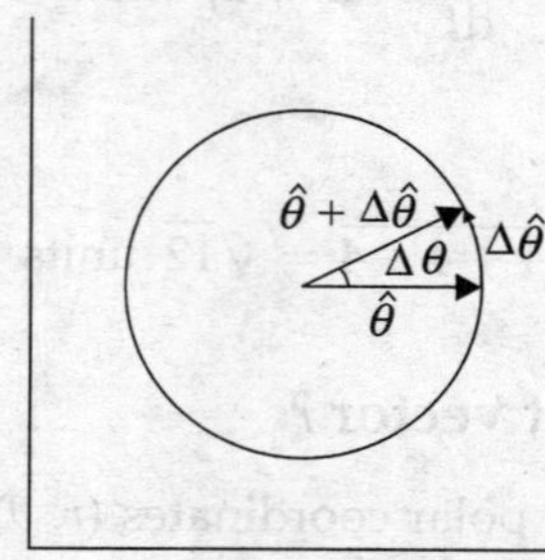

Figure 1.8: Change in the direction of unit vector $\hat{\theta}$.

as $|\hat{\theta}| = 1$. Thus, we have

$$\Delta\hat{\theta} = -\Delta\theta\,\hat{r} \qquad \text{and} \qquad \frac{\Delta\hat{\theta}}{\Delta t} = -\frac{\Delta\theta}{\Delta t}\,\hat{r}$$

Here, a negative sign appears because with the increase of θ, r decreases. For Δt tending to zero, we have

$$\lim_{\Delta t\to 0}\frac{\Delta\hat{\theta}}{\Delta t} = -\lim_{\Delta t\to 0}\frac{\Delta\theta}{\Delta t}\,\hat{r} \qquad \text{or} \qquad \frac{d\hat{\theta}}{dt} = -\frac{d\theta}{dt}\,\hat{r} \tag{1.10}$$

Equation (1.10) gives the rate of change of the unit vector $\hat{\theta}$. Notice that the rate of change of the unit vector $\hat{\theta}$ is opposite to the direction of the unit vector $\hat{r}$.

1.5.5 Curved motion of a body in a plane

When we consider a body moving in a plane along a curved path, we recognize that the velocity $\vec{v}$ at any instant is composed of the radial component $(dr/dt)\,\hat{r}$ and the transverse component $r d\hat{r}/dt = (r d\theta/dt)\,\hat{\theta}$. Thus, the expression for the velocity $\vec{v}$ is

$$\vec{v} = \frac{dr}{dt}\,\hat{r} + r\,\frac{d\theta}{dt}\,\hat{\theta} \tag{1.11}$$

For motion along a circle, the value of r does not change, and from equation (1.11), we have

$$\vec{v} = r\,\frac{d\theta}{dt}\,\hat{\theta} = r\omega\,\hat{\theta}$$

where $\omega = d\theta/dt$ is the angular velocity. Thus, we have

$$v = r\omega$$

1.5.6 Areal velocity

Suppose a body is moving along a curved path from a point P (r, θ) to another point Q $(r + \mathrm{d}r, \theta + \mathrm{d}\theta)$ in time $\mathrm{d}t$. The line joining the body and the centre O of motion sweeps an area and the rate of sweep of area is known as the areal velocity. The area swept by the line joining the body and the centre of motion is the area of the triangle OPQ, expressed as

$$\mathrm{d}A = \frac{1}{2}\,\mathrm{OP}\,\mathrm{PR}$$

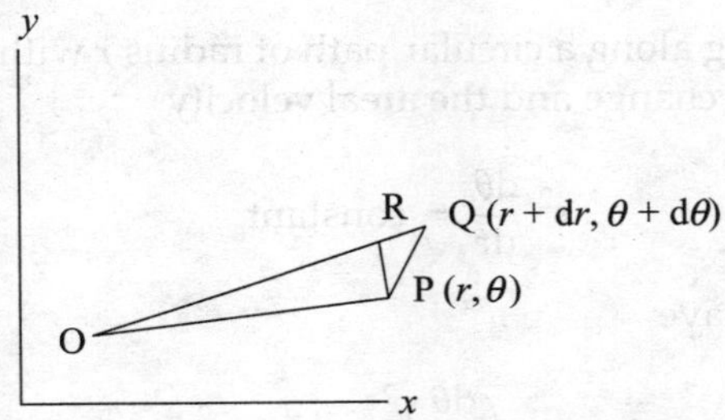

Here, PR is perpendicular from the point P on the line OQ. We have OP = r, PR = $r\,\mathrm{d}\theta$ and therefore

$$\mathrm{d}A = \frac{1}{2}\,r\,r\,\mathrm{d}\theta = \frac{1}{2}\,r^2\,\mathrm{d}\theta$$

Thus, the area swept per unit time, known as the areal velocity, is

$$\frac{\mathrm{d}A}{\mathrm{d}t} = \frac{1}{2}\,r^2\,\frac{\mathrm{d}\theta}{\mathrm{d}t} = \frac{1}{2}\,r^2\omega$$

where $\omega = \mathrm{d}\theta/\mathrm{d}t$ is the angular velocity. Kepler found that the areal velocity for a planet revolving around the sun is constant. We will discuss this in Chapter 4. The same law of constant areal velocity is applicable for the motion of a satellite revolving around its respective planet.

1.5.7 Acceleration for a curved motion

On differentiating equation (1.11) with respect to time t, we get

$$\frac{\mathrm{d}\vec{v}}{\mathrm{d}t} = \frac{\mathrm{d}^2r}{\mathrm{d}t^2}\,\hat{r} + \frac{\mathrm{d}r}{\mathrm{d}t}\,\frac{\mathrm{d}\hat{r}}{\mathrm{d}t} + \frac{\mathrm{d}r}{\mathrm{d}t}\,\frac{\mathrm{d}\theta}{\mathrm{d}t}\,\hat{\theta} + r\,\frac{\mathrm{d}^2\theta}{\mathrm{d}t^2}\,\hat{\theta} + r\,\frac{\mathrm{d}\theta}{\mathrm{d}t}\,\frac{\mathrm{d}\hat{\theta}}{\mathrm{d}t} \qquad (1.12)$$

Using equations (1.9) and (1.10) in (1.12), we have acceleration

$$\vec{a} = \frac{\mathrm{d}\vec{v}}{\mathrm{d}t} = \frac{\mathrm{d}^2r}{\mathrm{d}t^2}\,\hat{r} + \frac{\mathrm{d}r}{\mathrm{d}t}\,\frac{\mathrm{d}\theta}{\mathrm{d}t}\,\hat{\theta} + \frac{\mathrm{d}r}{\mathrm{d}t}\,\frac{\mathrm{d}\theta}{\mathrm{d}t}\,\hat{\theta} + r\,\frac{\mathrm{d}^2\theta}{\mathrm{d}t^2}\,\hat{\theta} - r\left(\frac{\mathrm{d}\theta}{\mathrm{d}t}\right)^2\hat{r}$$

By rearranging terms, we get the acceleration

$$\vec{a} = \left[\frac{\mathrm{d}^2r}{\mathrm{d}t^2} - r\left(\frac{\mathrm{d}\theta}{\mathrm{d}t}\right)^2\right]\hat{r} + \left[r\,\frac{\mathrm{d}^2\theta}{\mathrm{d}t^2} + 2\,\frac{\mathrm{d}r}{\mathrm{d}t}\frac{\mathrm{d}\theta}{\mathrm{d}t}\right]\hat{\theta}$$

$$= \left[\frac{\mathrm{d}^2 r}{\mathrm{d}t^2} - r\left(\frac{\mathrm{d}\theta}{\mathrm{d}t}\right)^2\right]\hat{r} + \frac{1}{r}\left[\frac{\mathrm{d}}{\mathrm{d}t}\left(r^2\frac{\mathrm{d}\theta}{\mathrm{d}t}\right)\right]\hat{\theta} = a_r\,\hat{r} + a_\theta\,\hat{\theta} \tag{1.13}$$

where

$$a_r = \frac{\mathrm{d}^2 r}{\mathrm{d}t^2} - r\left(\frac{\mathrm{d}\theta}{\mathrm{d}t}\right)^2 \qquad \text{and} \qquad a_\theta = \frac{1}{r}\left[\frac{\mathrm{d}}{\mathrm{d}t}\left(r^2\frac{\mathrm{d}\theta}{\mathrm{d}t}\right)\right]$$

are, respectively, known as the radial and angular components of acceleration in the plane polar coordinates (r, θ).

When a body is moving along a circular path of radius r with a constant angular velocity, the value of r does not change and the areal velocity

$$r^2\frac{\mathrm{d}\theta}{\mathrm{d}t} = \text{constant}$$

from equation (1.13), we have

$$\vec{a} = -r\left(\frac{\mathrm{d}\theta}{\mathrm{d}t}\right)^2\hat{r} = -r\omega^2\,\hat{r}$$

from equation (1.11), we have

$$\vec{v} = r\omega\,\hat{\theta}$$

Thus, the magnitude of acceleration for a circular motion is

$$a = r\omega^2 = \frac{v^2}{r}$$

1.6 Operator ∇

In this section we mention in brief the operator ∇, called the del operator. In mechanics, we generally deal with the Cartesian coordinates (x, y, z) and spherical polar coordinates (r, θ, ϕ). We will therefore give direct expressions for gradient, divergence, curl and Laplacian in Cartesian and spherical polar coordinate systems.

1.6.1 Cartesian coordinates

In the Cartesian coordinates (x, y, z), the operator ∇ is expressed as

$$\nabla \equiv \frac{\partial}{\partial x}\,\hat{\imath} + \frac{\partial}{\partial y}\,\hat{\jmath} + \frac{\partial}{\partial z}\,\hat{k}$$

where $\hat{\imath}$, $\hat{\jmath}$, $\hat{k}$ are the unit vectors along the x, y and z axes, respectively. The gradient of a scalar quantity ϕ is expressed as

$$\nabla\phi = \frac{\partial\phi}{\partial x}\,\hat{\imath} + \frac{\partial\phi}{\partial y}\,\hat{\jmath} + \frac{\partial\phi}{\partial z}\,\hat{k}$$

Thus, gradient of a scalar quantity is a vector quantity. The divergence of a vector quantity $\vec{A} = A_x\hat{i} + A_y\hat{j} + A_z\hat{k}$ is expressed as

$$\nabla . \vec{A} = \frac{\partial A_x}{\partial x} + \frac{\partial A_y}{\partial y} + \frac{\partial A_z}{\partial z}$$

Thus, divergence of a vector quantity is a scalar quantity. The curl of a vector quantity $\vec{A} = A_x\hat{i} + A_y\hat{j} + A_z\hat{k}$ is expressed as

$$\nabla \times \vec{A} = \begin{vmatrix} \hat{i} & \hat{j} & \hat{k} \\ \frac{\partial}{\partial x} & \frac{\partial}{\partial y} & \frac{\partial}{\partial z} \\ A_x & A_y & A_z \end{vmatrix}$$

$$= \left(\frac{\partial A_z}{\partial y} - \frac{\partial A_y}{\partial z}\right)\hat{i} + \left(\frac{\partial A_x}{\partial z} - \frac{\partial A_z}{\partial x}\right)\hat{j} + \left(\frac{\partial A_y}{\partial x} - \frac{\partial A_x}{\partial y}\right)\hat{k}$$

The Laplacian operator ∇^2 of a quantity Ψ is expressed as

$$\nabla^2\Psi = \frac{\partial^2\Psi}{\partial x^2} + \frac{\partial^2\Psi}{\partial y^2} + \frac{\partial^2\Psi}{\partial z^2}$$

When Ψ is a scalar quantity, then $\nabla^2\Psi$ is a scalar quantity and when Ψ is a vector quantity, then $\nabla^2\Psi$ is a vector quantity.

Exercise 1.22: We are given a scalar quantity $\phi = x^2y + y^2z + z^2x$ and a vector quantity $\vec{A} = x^2y\,\hat{i} + y^2z\,\hat{j} + z^2x\,\hat{k}$. Calculate (i) $\nabla\phi$, (ii) $\nabla . \vec{A}$ and (iii) $\nabla \times \vec{A}$.

Solution: (i) We have

$$\nabla\phi = \frac{\partial\phi}{\partial x}\,\hat{i} + \frac{\partial\phi}{\partial y}\,\hat{j} + \frac{\partial\phi}{\partial z}\,\hat{k}$$

$$= (2xy + z^2)\hat{i} + (2yz + x^2)\hat{j} + (2zx + y^2)\hat{k}$$

(ii) We have

$$\nabla . \vec{A} = \frac{\partial A_x}{\partial x} + \frac{\partial A_y}{\partial y} + \frac{\partial A_z}{\partial z}$$

$$= \frac{\partial(x^2y)}{\partial x} + \frac{\partial(y^2z)}{\partial y} + \frac{\partial(z^2x)}{\partial z} = 2xy + 2yz + 2zx$$

(iii) We have

$$\nabla \times \vec{A} = \begin{vmatrix} \hat{i} & \hat{j} & \hat{k} \\ \frac{\partial}{\partial x} & \frac{\partial}{\partial y} & \frac{\partial}{\partial z} \\ A_x & A_y & A_z \end{vmatrix} = \begin{vmatrix} \hat{i} & \hat{j} & \hat{k} \\ \frac{\partial}{\partial x} & \frac{\partial}{\partial y} & \frac{\partial}{\partial z} \\ x^2y & y^2z & z^2x \end{vmatrix}$$

$$= \left[\frac{\partial(z^2x)}{\partial y} - \frac{\partial(y^2z)}{\partial z}\right]\hat{i} + \left[\frac{\partial(x^2y)}{\partial z} - \frac{\partial(z^2x)}{\partial x}\right]\hat{j} + \left[\frac{\partial(y^2z)}{\partial x} - \frac{\partial(x^2y)}{\partial y}\right]\hat{k}$$

$$= -y^2\hat{i} - z^2\hat{j} - x^2\hat{k}$$

Exercise 1.23: For a scalar quantity ϕ and a vector quantity $\vec{A}$, show that (i) $\nabla \times (\nabla\phi) = 0$ and (ii) $\nabla.\,(\nabla\times \vec{A}) = 0$.

Solution: (i) We have

$$\nabla\phi = \frac{\partial\phi}{\partial x}\,\hat{i} + \frac{\partial\phi}{\partial y}\,\hat{j} + \frac{\partial\phi}{\partial z}\,\hat{k}$$

and

$$\nabla \times (\nabla\phi) = \begin{vmatrix} \hat{i} & \hat{j} & \hat{k} \\ \frac{\partial}{\partial x} & \frac{\partial}{\partial y} & \frac{\partial}{\partial z} \\ \frac{\partial\phi}{\partial x} & \frac{\partial\phi}{\partial y} & \frac{\partial\phi}{\partial z} \end{vmatrix}$$

$$= \left(\frac{\partial^2\phi}{\partial y\partial z} - \frac{\partial^2\phi}{\partial z\partial y}\right)\hat{i} - \left(\frac{\partial^2\phi}{\partial x\partial z} - \frac{\partial^2\phi}{\partial z\partial x}\right)\hat{j} + \left(\frac{\partial^2\phi}{\partial x\partial y} - \frac{\partial^2\phi}{\partial y\partial x}\right)\hat{k} = 0$$

(ii) We have

$$\nabla\times \vec{A} = \begin{vmatrix} \hat{i} & \hat{j} & \hat{k} \\ \frac{\partial}{\partial x} & \frac{\partial}{\partial y} & \frac{\partial}{\partial z} \\ A_x & A_y & A_z \end{vmatrix}$$

$$= \left(\frac{\partial A_z}{\partial y} - \frac{\partial A_y}{\partial z}\right)\hat{i} - \left(\frac{\partial A_z}{\partial x} - \frac{\partial A_x}{\partial z}\right)\hat{j} + \left(\frac{\partial A_y}{\partial x} - \frac{\partial A_x}{\partial y}\right)\hat{k}$$

and

$$\nabla.\,(\nabla\times \vec{A}) = \frac{\partial}{\partial x}\left(\frac{\partial A_z}{\partial y} - \frac{\partial A_y}{\partial z}\right) - \frac{\partial}{\partial y}\left(\frac{\partial A_z}{\partial x} - \frac{\partial A_x}{\partial z}\right) + \frac{\partial}{\partial z}\left(\frac{\partial A_y}{\partial x} - \frac{\partial A_x}{\partial y}\right)$$

$$= \left(\frac{\partial^2 A_z}{\partial x\partial y} - \frac{\partial^2 A_y}{\partial x\partial z}\right) - \left(\frac{\partial^2 A_z}{\partial y\partial x} - \frac{\partial^2 A_x}{\partial y\partial z}\right) + \left(\frac{\partial^2 A_y}{\partial z\partial x} - \frac{\partial^2 A_x}{\partial z\partial y}\right) = 0$$

1.6.2 Spherical polar coordinates

Spherical polar coordinates are preferred while dealing with a system having spherical shape. Example of such system is the hydrogen atom. In the spherical polar coordinates (r, θ, ϕ), gradient of a scalar quantity ψ is expressed as

$$\nabla\psi = \frac{\partial\psi}{\partial r}\,\hat{r} + \frac{1}{r}\frac{\partial\psi}{\partial\theta}\,\hat{\theta} + \frac{1}{r\sin\theta}\frac{\partial\psi}{\partial\phi}\,\hat{\phi}$$

where $\hat{r}$, $\hat{\theta}$, $\hat{\phi}$ are unit vectors corresponding to the coordinates r, θ, ϕ, respectively. Let us consider a vector quantity

$$\vec{A} = A_r\,\hat{r} + A_\theta\,\hat{\theta} + A_\phi\,\hat{\phi}$$

Here, A_r, A_θ, A_ϕ are, respectively, known as r, θ, ϕ components of the vector $\vec{A}$. The divergence of $\vec{A}$ is expressed as

$$\nabla.\,\vec{A} = \frac{1}{r^2\sin\theta}\left[\frac{\partial}{\partial r}(A_r r^2\sin\theta) + \frac{\partial}{\partial\theta}(A_\theta r\sin\theta) + \frac{\partial}{\partial\phi}(A_\phi r)\right]$$

The curl of $\vec{A}$ is expressed as

$$\nabla\times\vec{A} = \frac{1}{r^2\sin\theta}\begin{bmatrix}\hat{r} & r\,\hat{\theta} & r\sin\theta\,\hat{\phi}\\ \frac{\partial}{\partial r} & \frac{\partial}{\partial\theta} & \frac{\partial}{\partial\phi}\\ A_r & rA_\theta & r\sin\theta A_\phi\end{bmatrix}$$

$$= \frac{1}{r^2\sin\theta}\left[\hat{r}\left\{\frac{\partial(r\sin\theta A_\phi)}{\partial\theta} - \frac{\partial(rA_\theta)}{\partial\phi}\right\} - r\,\hat{\theta}\left\{\frac{\partial(r\sin\theta A_\phi)}{\partial r} - \frac{\partial(rA_\theta)}{\partial\phi}\right\}\right.$$
$$\left. + r\sin\theta\,\hat{\phi}\left\{\frac{\partial(rA_\theta)}{\partial r} - \frac{\partial A_r}{\partial\theta}\right\}\right]$$

The Laplacian operator of a quantity Ψ is expressed as

$$\nabla^2\Psi = \frac{1}{r^2}\left[\frac{\partial}{\partial r}\left(r^2\,\frac{\partial\Psi}{\partial r}\right) + \frac{1}{\sin\theta}\frac{\partial}{\partial\theta}\left(\sin\theta\,\frac{\partial\Psi}{\partial\theta}\right) + \frac{1}{\sin^2\theta}\frac{\partial^2\Psi}{\partial\phi^2}\right]$$

When Ψ is a scalar quantity, then $\nabla^2\Psi$ is a scalar quantity and when Ψ is a vector quantity, $\nabla^2\Psi$ is a vector quantity.

Multiple-choice questions

1. Which of the following relations is not satisfied by vectors?

(i) $\vec{A}\cdot\vec{B} = \vec{B}\cdot\vec{A}$ (ii) $\vec{A}\times\vec{B} = \vec{B}\times\vec{A}$

(iii) $\vec{A}\,.(\vec{B}\times\vec{C}) = (\vec{A}\times\vec{B}).\,\vec{C}$ (iv) $\vec{A} + (\vec{B}+\vec{C}) = (\vec{A}+\vec{B}) + \vec{C}$

Ans. (ii)

2. Four forces $\vec{F}_1$, $\vec{F}_2$, $\vec{F}_3$, $\vec{F}_4$ acting on a body are in equilibrium. When $\vec{F}_1 = 2\hat{\imath} + 3\hat{\jmath} - 4\hat{k}$, $\vec{F}_2 = 3\hat{\imath} - 2\hat{\jmath} + \hat{k}$, $\vec{F}_3 = \hat{\imath} + \hat{\jmath} - \hat{k}$, the fourth force $\vec{F}_4$ is

(i) $6\hat{\imath} + 2\hat{\jmath} - 4\hat{k}$ (ii) $4\hat{\imath} - 2\hat{k}$

(iii) $2\hat{\imath} - 4\hat{\jmath} + 4\hat{k}$ (iv) $-6\hat{\imath} - 2\hat{\jmath} + 4\hat{k}$

Ans. (iv)

3. Which of the following is not a vector quantity?

(i) Displacement (ii) Velocity

(iii) Work (iv) Momentum

Ans. (iii)

4. For unit vectors $\hat{i}$, $\hat{j}$, $\hat{k}$ in Cartesian coordinates, which of the following relations is wrong?

(i) $\hat{i}.\hat{j} = \hat{k}$ (ii) $\hat{i} \times \hat{j} = \hat{k}$

(iii) $\hat{i}.\hat{i} = 1$ (iv) $\hat{i} \times \hat{i} = 0$

Ans. (i)

5. When $\vec{A} = 2\hat{i} + 2\hat{j} - \hat{k}$ and $\vec{B} = \hat{i} + 2\hat{j} - 3\hat{k}$, the $\vec{A} \times \vec{B}$ is

(i) $-4\hat{i} - 5\hat{j} + 2\hat{k}$ (ii) $-4\hat{i} + 5\hat{j} + 2\hat{k}$

(iii) $2\hat{i} + 4\hat{j} + 6\hat{k}$ (iv) $4\hat{i} - 5\hat{j} - 2\hat{k}$

Ans. (ii)

6. A particle having a charge q is moving in a magnetic field $\vec{B}$ with velocity $\vec{v}$. The force acting on the particle is

(i) $q\,\vec{B} \times \vec{v}$ (ii) $q\,\vec{v}\,.\,\vec{B}$

(iii) $q\,\vec{v} \times \vec{B}$ (iv) $q(\vec{v} + \vec{B})$

Ans. (iii)

7. Which of the following relations is wrong?

(i) $\vec{A} \times (\vec{B} \times \vec{C}) = \vec{B}\,(\vec{A}\,.\,\vec{C}) - \vec{C}\,(\vec{A}\,.\,\vec{B})$ (ii) $\vec{A} \times \vec{B} = \vec{B} \times \vec{A}$

(iii) $\vec{A}\,.\,\vec{B} = \vec{B}\,.\,\vec{A}$ (iv) $\vec{A}\,.(\vec{B} \times \vec{C}) = (\vec{A} \times \vec{B}).\,\vec{C}$

Ans. (ii)

8. For scalar quantities ϕ and λ, and vector quantity $\vec{A}$, which of the following relations is possible?

(i) $\nabla\phi = \lambda\phi$ (ii) $\nabla.\,\vec{A} = \lambda\,\vec{A}$

(iii) $\nabla \times \vec{A} = \lambda\,\vec{A}$ (iv) None of them

Ans. (iii)

9. For a scalar quantity ϕ and vector quantity $\vec{A}$, which of the following relations is wrong?

(i) $\nabla \times (\nabla \phi) = 0$ (ii) $\nabla.(\nabla \times \vec{A}) = 0$

(iii) $\nabla \times (\nabla \times \vec{A}) = 0$ (iv) $\nabla.(\nabla \phi) = \nabla^2 \phi$

Ans. (iii)

10. A force which does not do work is
    (i) Gravitational force (ii) Lorentz force
    (iii) Electric force (iv) None of them

Ans. (ii)

Exercises

1. What is the difference between a scalar quantity and a vector quantity? Give two examples of each of them.

2. Define addition of two vectors. Can two vectors having different dimensions be added? With the help of two vectors, prove that addition of two vectors is commutative.

3. Define subtraction of one vector from the other. Suppose $\vec{A}$ and $\vec{B}$ are two vectors having different dimensions. Can we calculate $(\vec{A} - \vec{B})$? With the help of two vectors, prove that subtraction is not commutative.

4. Show that addition of two vectors is commutative whereas subtraction is not.

5. For two vectors $\vec{A} = A_x\hat{\imath} + A_y\hat{\jmath} + A_z\hat{k}$ and $\vec{B} = B_x\hat{\imath} + B_y\hat{\jmath} + B_z\hat{k}$, show that $k(\vec{A} + \vec{B}) = k\vec{A} + k\vec{B}$.

6. For two vectors $\vec{A}$ and $\vec{B}$, show that $\vec{A} + \vec{B} = \vec{B} + \vec{A}$.

7. For three vectors $\vec{A}, \vec{B}, \vec{C}$, show that $\vec{A} + (\vec{B} + \vec{C}) = (\vec{A} + \vec{B}) + \vec{C}$.

8. For two vectors $\vec{A} = 2\hat{\imath} + 3\hat{\jmath} + 2\hat{k}$ and $\vec{B} = -\hat{\imath} + 2\hat{\jmath} + 3\hat{k}$, calculate $2\vec{A} - \vec{B}$ and $2\vec{A} + 3\vec{B}$.

9. Define velocity and acceleration of a body. The displacement of a body is expressed as

$$\vec{r} = (t^2 + 2)\hat{\imath} + (t + 2)^2\hat{\jmath} + (t - 1)^2\hat{k}$$

where t denotes the time. Calculate at $t = 2$ (i) magnitude of displacement, (ii) velocity and its magnitude and (iii) acceleration and its magnitude.

10. Define the equilibrium condition for a set of forces working simultaneously. Under the simultaneous action of four forces, a body is in equilibrium. When the three forces are $\vec{F}_1 = 2\hat{\imath} - 2\hat{\jmath} + 3\hat{k}$, $\vec{F}_2 = 2\hat{\imath} - 2\hat{\jmath} + \hat{k}$ and $\vec{F}_3 = 3\hat{\imath} - 3\hat{\jmath} + 4\hat{k}$, find out the fourth force.

11. Define scalar multiplication of two vectors. With the help of two vectors, prove that scalar multiplication is commutative.

12. Suppose $\vec{A}$ and $\vec{B}$ are two vectors having different dimensions. Can we calculate $(\vec{A} \cdot \vec{B})$? Can we calculate $(\vec{A} \times \vec{B})$?

13. Write an expression for the force working on a charged particle placed in an electric field. On application of an electric field $\vec{E} = 4\hat{\imath} + 3\hat{\jmath} + 2\hat{k}$, a body carrying a charge of 2 C is displaced from a position $(-1, 3, 4)$ to other point $(3, -2, 6)$. Calculate the amount of work done.

14. A magnetic field $\vec{B} = 3\hat{\imath} - \hat{\jmath} + 4\hat{k}$ is applied on a particle carrying a charge of 2 C and moving with the velocity $\vec{v} = \hat{\imath} + 3\hat{\jmath} - 4\hat{k}$. Show that the work done by the magnetic field is zero.

15. Obtain the angle between the vectors $\vec{A} = 4\hat{\imath} + 2\hat{\jmath} + \hat{k}$ and $\vec{B} = 2\hat{\imath} - \hat{\jmath} + 2\hat{k}$.

16. Define vector multiplication of two vectors. With the help of two vectors, prove that vector multiplication is not commutative.

17. Coordinates of points A, B, C in the following parallelogram are (2, 1, 3), (1, 4, 5), (1, −2, 3), respectively. Calculate the area of the parallelogram.

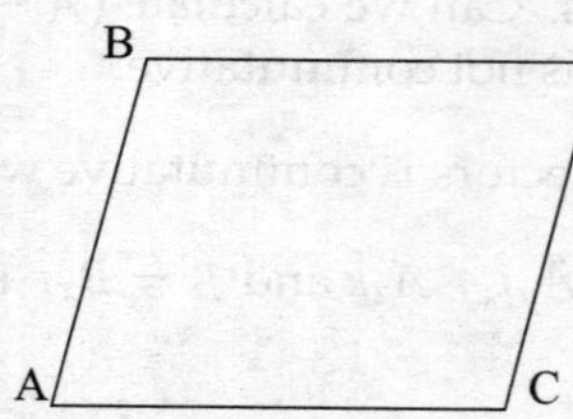

18. Define the scalar triple product of three vectors. For three vectors, $\vec{A} = 2\hat{\imath} - 2\hat{\jmath} + 5\hat{k}$, $\vec{B} = -\hat{\imath} + 2\hat{\jmath} - 3\hat{k}$, $\vec{C} = 3\hat{\imath} + \hat{\jmath} + 4\hat{k}$, calculate $\vec{A} \cdot (\vec{B} \times \vec{C})$.

19. Define the vector triple product of three vectors. For three vectors, $\vec{A} = \hat{\imath} + 2\hat{\jmath} + 3\hat{k}$, $\vec{B} = \hat{\imath} - \hat{\jmath} + 3\hat{k}$, $\vec{C} = 3\hat{\imath} - \hat{\jmath} - 4\hat{k}$, calculate $\vec{A} \times (\vec{B} \times \vec{C})$.

20. For three vectors $\vec{A}$, $\vec{B}$, $\vec{C}$, show that

$$\vec{A} \times (\vec{B} \times \vec{C}) = \vec{B}\,(\vec{A} \cdot \vec{C}) - \vec{C}\,(\vec{A} \cdot \vec{B})$$

21. A body of mass 2 kg is moving with a velocity $\vec{v} = 4\hat{\imath} - 2\hat{\jmath} + 3\hat{k}$ when it is at a point P whose coordinates are (2, 1, 3). Calculate the angular momentum of the body about an axis passing through a point Q whose coordinates are (−2, 0, 4).

22. Coordinates of the points A, B, C, D in the following parallelepiped are (1, 2, 3), (2, 4, 5), (−1, 0, −2), (1, −1, 3), respectively. Calculate volume of the parallelepiped.

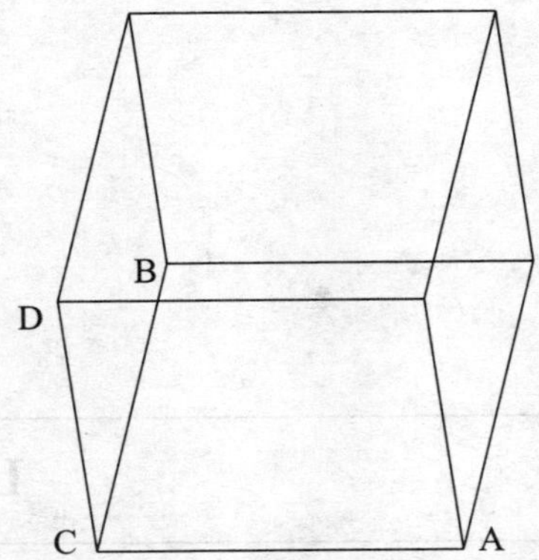

23. Write the expressions for gradient of a scalar quantity ϕ, divergence of a vector quantity $\vec{A}$, curl of a vector quantity $\vec{A}$ in Cartesian coordinates (x, y, z).

24. For a scalar quantity ϕ and a vector quantity $\vec{A}$, show that (i) $\nabla \times (\nabla\phi) = 0$ and (ii) $\nabla.(\nabla\times\vec{A}) = 0$.

25. For $\vec{A} = x^2y\hat{i} + y^2z\hat{j} + z^2x\hat{k}$, prove that $\nabla.(\nabla\times\vec{A}) = 0$.

26. For $\phi = x^2y^2z^2$, prove that $\nabla \times (\nabla\phi) = 0$.

27. Write expressions for gradient of a scalar quantity ψ, divergence of a vector quantity $\vec{A}$, curl of a vector quantity $\vec{A}$ in spherical polar coordinates (r, θ, ϕ).

28. Write notes on the following:
    (i) Addition of two vectors expressed in Cartesian coordinates
    (ii) Subtraction of one vector from the other when the vectors are expressed in Cartesian coordinates
    (iii) Velocity and acceleration in two-dimensional polar coordinates
    (iv) Multiplication of two vectors expressed in Cartesian coordinates
    (v) Work in terms of vectors
    (vi) Angular momentum in terms of vectors
    (vii) Torque in terms of vectors
    (viii) Lorentz force in terms of vectors
    (ix) Velocity of a body in the plane polar coordinates
    (x) Acceleration of a body in the plane polar coordinates
    (xi) Velocity and acceleration of a body moving along a circular path
    (xii) Sine law
    (xiii) Cosine law
    (xiv) Areal velocity
    (xv) Scalar triple product of vectors
    (xvi) Vector triple product of vectors

CHAPTER 2

LAWS OF MOTION

Mechanics is a branch of science and technology where we deal with various kinds of motions of bodies. Based on the observations for the motions of bodies, Newton put forth three laws, often called the Newton's laws of motion. In the present chapter, we shall discuss about them. While discussing the Newton's laws of motion, we shall assume that the atmosphere around us is not imposing any force on the moving bodies. That is, we shall assume that the atmosphere is not present. For expressing the motion, we require some reference frame and a coordinate system.

2.1 Inertial reference frame

For describing motion of a body, we require a reference frame. Newton considered an absolute reference frame and a universal time. For an absolute reference frame, its origin as well as its axes are absolutely fixed. Distances (displacements) of bodies are measured with reference to this reference frame.

The reference frame in which Newton's laws of motion are valid is known as an inertial reference frame. In fact such a reference frame does not exist, as no material body is at absolute rest and therefore we cannot have an absolute reference frame. Even a reference frame attached to the earth is not an absolute one, as the earth revolves around the sun, as well as it spins about its own axis. Further, sun also moves in our galaxy, the Milky Way. In fact, no body in the universe is at the absolute rest.

Let us imagine an absolute reference frame. Suppose, a reference frame S_1 is moving with a uniform linear velocity $\vec{v}_1$ relative to the absolute reference frame. When the velocity of the body is linear uniform, we know from the Newton's first law of motion that there is no external force acting on the body. The body is placed at rest in this reference frame. Now, we have another reference frame S_2 moving with a uniform linear velocity $\vec{v}_2$ relative to the absolute reference frame. The reference frame S_2 is moving with a uniform linear velocity $(\vec{v}_2 - \vec{v}_1)$ relative to the reference frame S_1. Although the velocity $\vec{v}_1$ of S_1 cannot be defined without reference to the absolute reference frame, the velocity of S_2 relative to S_1 can be defined. Such reference frames which are moving with the uniform linear relative velocities are known as the inertial reference frames. We may now say that an inertial reference frame is that in which Newton's laws of motion are valid.

An accelerated reference frame cannot be considered as an inertial reference frame. In a rotational motion, a particle is accelerated as the direction of its motion goes on changing. Owing to the revolution and spin motions of the earth, any reference frame attached to the earth cannot be taken as an inertial reference frame. However, for practical purposes, a reference frame attached to the earth or moving with a uniform velocity relative to the earth is taken as an inertial reference frame for the following reasons.

2.1.1 Centrifugal acceleration

While revolving around the sun, earth completes one round in 1 year. Thus, the revolution can be taken as almost linear. In the spin motion, the earth rotates through an angle of 2π rad in 24 hr. Hence, the angular velocity of the earth is

$$\omega = \frac{2\pi}{24 \times 60 \times 60} \approx 7.3 \times 10^{-5} \text{ rad/s}$$

On surface of the earth, the centrifugal acceleration is maximum at its equator. The radius of the earth is approximately 6.4×10^6 m. Thus, the centrifugal acceleration at the equator is

$$\omega^2 r = (7.3 \times 10^{-5})^2 \times 6.4 \times 10^6 = 0.034 \text{ m/s}^2$$

The acceleration due to gravity on the earth's surface is 9.81 m/s^2. Thus, the maximum value of the centrifugal acceleration is only 0.3% of the acceleration due to gravity. The centrifugal acceleration is very small and can be neglected in comparison to the acceleration due to gravity.

2.1.2 Coriolis acceleration

Another contribution towards non-inertial reference frames comes from the Coriolis acceleration, which depends on the velocity of the particle moving relative to the earth and its magnitude is $2\omega v$. For a velocity of 1 km/s (i.e., 3600 km/hr), which is quite high, we have the Coriolis acceleration as

$$2\omega v = 2(7.3 \times 10^{-5}) \times 10^3 = 0.146 \text{ m/s}^2$$

Thus, the Coriolis acceleration is only 1.49% of the acceleration due to gravity. The Coriolis acceleration is very small and can be neglected in comparison to the acceleration due to gravity.

Since both the centrifugal acceleration as well as Coriolis acceleration are much smaller than the acceleration due to gravity, therefore, the earth for practical purposes can be treated as an inertial reference frame.

2.1.3 Non-inertial reference frame

A reference frame which is not inertial belong to the category of non-inertial reference frame. That is, a reference frame which is accelerated or rotating is known as non-inertial reference frame. Non-inertial reference frames are discussed in detail in Chapter 7.

2.1.4 Conservative and non-conservative forces

A force which depends on the coordinates and not on the velocities of the particles of the system is known as the 'conservative force'. Examples, of such forces are the gravitational force between two masses, Coulomb force between two charges, restoring force, etc. Obviously, in the conservative field, potential energy will also depend on the coordinates only.

Contrary to the conservative forces, a non-conservative force depends on the velocities of the particles besides on their coordinates. Examples of such forces are the Lorentz force, frictional force, etc.

2.2 Galilean transformations

Let us consider two inertial reference frames S and S′, as shown in Figure 2.1. For convenience, suppose S′ is moving with a constant linear velocity V along x-axis with respect to S. At the time $t = 0$ both S and S′ coincided with each other. At time t, the distance OO′ is Vt; and the coordinates of P measured in S and S′ are (x, y, z) and (x', y', z'), respectively. Then, we have

$$x' = x - Vt \qquad y' = y \qquad z' = z \tag{2.1}$$

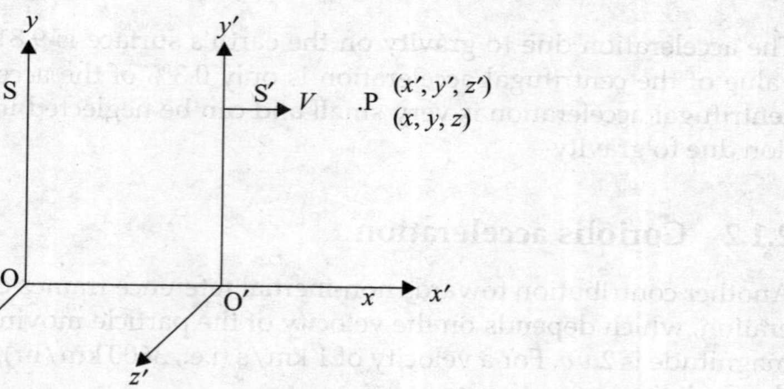

Figure 2.1: For convenience, let the reference frame S′ be moving with velocity V along x-axis.

In the classical treatment, the time is taken as absolute, that is the time does not depend on the reference frame. It is equivalent to say that the time and space are independent of each other, or that there is no coupling between the space and time. The inverse transformations are

$$x = x' + Vt \qquad y = y' \qquad z = z' \tag{2.2}$$

The equations (2.1) and (2.2) are known as the Galilean transformations for the coordinates. Remember that here we have assumed the motion of S′ along the x-axis. Differentiating equation (2.1) with respect to t, we have

$$\frac{dx}{dt} = \frac{dx'}{dt} - V \qquad \frac{dy}{dt} = \frac{dy'}{dt} \qquad \frac{dz}{dt} = \frac{dz'}{dt} \tag{2.3}$$

The velocity $\vec{v}$ in S is

$$\vec{v} = \frac{dx}{dt}\,\hat{i} + \frac{dy}{dt}\,\hat{j} + \frac{dz}{dt}\,\hat{k} \tag{2.4}$$

Using equation (2.3) in (2.4), we have

$$\vec{v} = \left[\frac{dx'}{dt} - V\right]\hat{i} + \frac{dy'}{dt}\,\hat{j} + \frac{dz'}{dt}\,\hat{k} \tag{2.5}$$

or

$$\vec{v} = \vec{v}' - \vec{V} \tag{2.6}$$

where

$$\vec{v'} = \frac{dx'}{dt}\,\hat{i} + \frac{dy'}{dt}\,\hat{j} + \frac{dz'}{dt}\,\hat{k}$$

is the velocity in S′. Differentiating equation (2.2) with respect to t, we have

$$\frac{dx'}{dt} = \frac{dx}{dt} + V \qquad \frac{dy'}{dt} = \frac{dy}{dt} \qquad \frac{dz'}{dt} = \frac{dz}{dt} \tag{2.7}$$

The velocity $\vec{v}'$ in S′ is

$$\vec{v}' = \frac{dx'}{dt}\,\hat{i} + \frac{dy'}{dt}\,\hat{j} + \frac{dz'}{dt}\,\hat{k} \tag{2.8}$$

Using equation (2.7) in (2.8), we have

$$\vec{v'} = \left[\frac{dx}{dt} + V\right]\hat{i} + \frac{dy}{dt}\,\hat{j} + \frac{dz}{dt}\,\hat{k}$$

or

$$\vec{v'} = \vec{v} + \vec{V} \tag{2.9}$$

where

$$\vec{v} = \frac{dx}{dt}\,\hat{i} + \frac{dy}{dt}\,\hat{j} + \frac{dz}{dt}\,\hat{k}$$

is the velocity in S. Equations (2.6) and (2.9) are known as the Galilean transformations for velocity. Differentiating equation (2.3) with respect to t, we have

$$\frac{d^2x}{dt^2} = \frac{d^2x'}{dt^2} \qquad \frac{d^2y}{dt^2} = \frac{d^2y'}{dt^2} \qquad \frac{d^2z}{dt^2} = \frac{d^2z'}{dt^2} \tag{2.10}$$

The acceleration $\vec{a}$ in S is

$$\vec{a} = \frac{d^2x}{dt^2}\,\hat{i} + \frac{d^2y}{dt^2}\,\hat{j} + \frac{d^2z}{dt^2}\,\hat{k} \tag{2.11}$$

Using equation (2.11) in (2.10), we have

$$\vec{a} = \frac{d^2x'}{dt^2}\,\hat{i} + \frac{d^2y'}{dt^2}\,\hat{j} + \frac{d^2z'}{dt^2}\,\hat{k} \qquad \text{or} \qquad \vec{a} = \vec{a'}$$

where

$$\vec{a'} = \frac{d^2x'}{dt^2}\,\hat{i} + \frac{d^2y'}{dt^2}\,\hat{j} + \frac{d^2z'}{dt^2}\,\hat{k}$$

is the acceleration in S′. Differentiating equation (2.3) with respect to t, we have

$$\frac{d^2x'}{dt^2} = \frac{d^2x}{dt^2} \qquad \frac{d^2y'}{dt^2} = \frac{d^2y}{dt^2} \qquad \frac{d^2z'}{dt^2} = \frac{d^2z}{dt^2} \tag{2.12}$$

The acceleration $\vec{a'}$ in S′ is

$$\vec{a'} = \frac{d^2x'}{dt^2}\,\hat{i} + \frac{d^2y'}{dt^2}\,\hat{j} + \frac{d^2z'}{dt^2}\,\hat{k} \tag{2.13}$$

Using equation (2.12) in (2.13), we have

$$\vec{a'} = \frac{d^2x}{dt^2}\,\hat{i} + \frac{d^2y}{dt^2}\,\hat{j} + \frac{d^2z}{dt^2}\,\hat{k} \qquad \text{or} \qquad \vec{a'} = \vec{a}$$

where

$$\vec{a} = \frac{d^2x}{dt^2}\,\hat{i} + \frac{d^2y}{dt^2}\,\hat{j} + \frac{d^2z}{dt^2}\,\hat{k}$$

is the acceleration in S. Thus, under the Lorentz transformations, acceleration of a body is independent of the reference frame. That is, the acceleration is invariant under Lorentz transformations.

2.3 Coordinate systems

We generally deal with two types of reference frames: (i) laboratory coordinate system and (ii) centre-of-mass system. Each of them has its own importance.

2.3.1 Laboratory coordinate system

This reference frame may be a one which is attached to a laboratory situated on the surface of the earth or is moving relative to the earth. Such a reference frame is known as the laboratory coordinate system. Positions of various bodies are measured relative to the axes of the coordinate system. For example, the coordinates of the point P in the Cartesian coordinates system are (x, y, z). When we use spherical polar coordinates system, the coordinates of the point are (r, θ, ϕ). Similarly, we can express coordinates of a point in other coordinate systems.

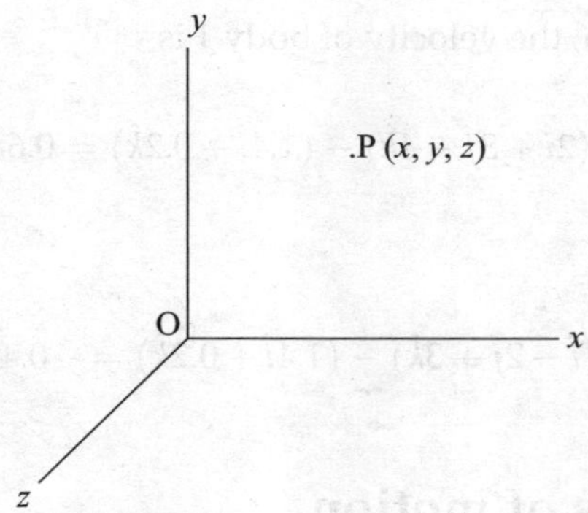

2.3.2 Centre-of-mass system

In some situations, for convenience, we use the centre-of-mass system in which the centre of mass is taken to be at rest. A centre-of-mass system is that in which the velocities of bodies are measured relative to the velocity of the centre of mass of the system. For a system of masses $m_1, m_2, \ldots, m_n$ moving with velocities $\vec{v}_1, \vec{v}_2, \ldots, \vec{v}_n$, respectively, the velocity $\vec{V}$ of the centre of mass is expressed as

$$\vec{V} = \frac{m_1\vec{v}_1 + m_2\vec{v}_2 + \ldots + m_n\vec{v}_n}{m_1 + m_2 + \ldots + m_n}$$

Since velocities in the centre-of-mass system are measured relative to the velocity of the centre of mass of the system, the velocities of the bodies in the centre-of-mass system are

$$\vec{u}_1 = \vec{v}_1 - \vec{V}, \qquad \vec{u}_2 = \vec{v}_2 - \vec{V}, \qquad \ldots, \qquad \vec{u}_n = \vec{v}_n - \vec{V}$$

respectively. It is not necessary that the centre of mass of the system is stationary. When the centre of mass is moving with a constant velocity $\vec{V}$, the displacement $\vec{r}$ of the centre of mass in time t is

$$\vec{r} = \vec{V}\,t$$

Exercise 2.1: Two bodies 1 and 2 of masses 0.2 kg and 0.3 kg are moving with velocities $\vec{v}_1 = 2\hat{\imath} + 3\hat{\jmath} - 4\hat{k}$ and $\vec{v}_2 = \hat{\imath} - 2\hat{\jmath} + 3\hat{k}$, respectively, in a laboratory system of reference. Calculate the velocities of bodies in the centre-of-mass system.

Solution: For the bodies 1 and 2, we have $m_1 = 0.2$ kg and $m_2 = 0.3$ kg, respectively. The velocity of the centre of mass is

$$\vec{V} = \frac{m_1\vec{v}_1 + m_2\vec{v}_2}{m_1 + m_2}$$

$$= \frac{0.2(2\hat{\imath} + 3\hat{\jmath} - 4\hat{k}) + 0.3(\hat{\imath} - 2\hat{\jmath} + 3\hat{k})}{0.2 + 0.3} = 1.4\hat{\imath} + 0.2\hat{k}$$

In the centre-of-mass system, the velocity of body 1 is

$$\vec{u}_1 = \vec{v}_1 - \vec{V} = (2\hat{i} + 3\hat{j} - 4\hat{k}) - (1.4\hat{i} + 0.2\hat{k}) = 0.6\hat{i} + 3\hat{j} - 4.2\hat{k}$$

and the velocity of body 2 is

$$\vec{u}_2 = \vec{v}_2 - \vec{V} = (\hat{i} - 2\hat{j} + 3\hat{k}) - (1.4\hat{i} + 0.2\hat{k}) = -0.4\hat{i} - 2\hat{j} + 2.8\hat{k}$$

2.4 Newton's laws of motion

Based on the observations of motions of bodies, Newton propounded three laws of motion, also known as Newton's laws of motion:

(i) Law of inertia

(ii) Law of motion

(ii) Law of action and reaction

Let us discuss about them one by one.

2.4.1 Law of inertia

This is also known as Newton's first law of motion. This law states that a body continues to remain in its state of motion or rest until and unless some external force is applied on it. That is, in absence of an external force, when a body is at rest, it will continue to remain at rest and when a body is moving, it will continue to move with the same velocity. We can say that the acceleration $\vec{a}$ of a body is zero when no external force $\vec{F}$ is applied on it. That is

$$\vec{a} = 0 \qquad \text{when} \qquad \vec{F} = 0$$

2.4.2 Law of motion

This is also known as Newton's second law of motion. This law states that the rate of change of momentum of a body is equal to the external force applied on it. The applied force may be of any kind. For example, mechanical force, electric force, magnetic force, etc. For a body of mass m moving with velocity $\vec{v}$ the momentum is $\vec{p} = m\vec{v}$. From the second law of motion, we have

$$\vec{F} = \frac{d\vec{p}}{dt} \qquad \text{or} \qquad \vec{F} = \frac{d(m\vec{v})}{dt} \tag{2.14}$$

For all motions in our day-to-day life, the mass of a body is constant.[1] Thus, we have

$$\vec{F} = m\frac{d\vec{v}}{dt} = m\vec{a}$$

where $\vec{a} = d\vec{v}/dt$ is the acceleration of the body. It shows that after application of an external force, the body is accelerated. The acceleration may be positive or negative, depending on the direction of the force applied on the body.

Note: Equation (2.14) is considered to be the basic equation for all the Newton's laws of motion. From this equation, we can also discuss Newton's first law of motion, that is, when $\vec{F} = 0$, we have $\vec{a} = 0$.

2.4.3 Law of action and reaction

This is also known as Newton's third law of motion. For this law, we require at least two bodies. This law states that whenever two particles interact, the force $\vec{F}_{21}$ exerted by the particle 1 on particle 2 is equal and opposite to the force $\vec{F}_{12}$ exerted by the particle 2 on the particle 1. That is

$$\vec{F}_{12} = -\vec{F}_{21} \qquad \text{or} \qquad \vec{F}_{12} + \vec{F}_{21} = 0$$

We shall see that this law is a basis for the conservation of momentum.

Conservation of momentum

Let us consider two particles 1 and 2 interacting with each other. Let $\vec{p}_1$ and $\vec{p}_2$ be momenta of the particles 1 and 2, respectively. Suppose $\vec{F}_{21}$ is the force exerted by the particle 1 on particle 2 and $\vec{F}_{12}$ is the force exerted by the particle 2 on particle 1. Then, we have

$$\frac{d\vec{p}_1}{dt} = \vec{F}_{12} \qquad \text{and} \qquad \frac{d\vec{p}_2}{dt} = \vec{F}_{21} \tag{2.15}$$

For Newton's third law of motion, we have

$$\vec{F}_{12} = -\vec{F}_{21} \qquad \text{or} \qquad \vec{F}_{12} + \vec{F}_{21} = 0 \tag{2.16}$$

Using equation (2.15) in (2.16), we have

$$\frac{d\vec{p}_1}{dt} + \frac{d\vec{p}_1}{dt} = 0 \qquad \text{or} \qquad \frac{d}{dt}(\vec{p}_1 + \vec{p}_2) = 0$$

[1]We know about the variation of mass of a body moving with velocity V as

$$m = \frac{m_0}{\sqrt{1 - V^2/c^2}}$$

where m_0 is the rest mass of the body. In our day-to-day life, velocities of all motion are much smaller than the speed of light ($V << c$) and therefore, V^2/c^2 can be neglected in comparison to 1 and $m = m_0$. That is, the mass of a body is constant.

It shows that $(\vec{p}_1 + \vec{p}_2)$ = constant. That is the momentum of the system is conserved. We can consider a system of two particles 1 and 2, where no external force is applied on the system. Then the rate of change of total momentum of the system is zero. This conclusion can be derived from Newton's second law of motion applied to the system of two particles 1 and 2.

2.5 Equation of motion

For a particle of mass m moving with velocity $\vec{v}$, Newton's second law of motion gives

$$\vec{F} = \frac{d(m\vec{v})}{dt} = m\frac{d\vec{v}}{dt} = m\frac{d^2\vec{r}}{dt^2}$$

This is known as the equation of motion. Here, the external force $\vec{F}$ may be of any kind. When $\vec{r} = x\hat{i} + y\hat{j} + z\hat{k}$, we have

$$\vec{F} = m\left[\frac{d^2x}{dt^2}\hat{i} + \frac{d^2y}{dt^2}\hat{j} + \frac{d^2z}{dt^2}\hat{k}\right]$$

By knowing the force $\vec{F}$, we can find out velocity and position of a particle at time t.

Exercise 2.2: Show that for a particle moving along a circular (curved) path, the force acting on it can never be zero.

Solution: Suppose a particle of mass m is moving with velocity $\vec{v}$. For the force $\vec{F}$ acting on the particle, Newton's second law of motion gives

$$\vec{F} = m\frac{d\vec{v}}{dt}$$

For a particle moving along a circular (curved) path, the direction of velocity changes and therefore the velocity, which has magnitude as well as direction, changes. Since the velocity is changing with time, the force $\vec{F}$ acting on the body cannot be zero.

Exercise 2.3: Show that for a particle moving with constant velocity $\vec{v}_0$, the position $\vec{r}$ of the particle at time t is $\vec{r} = \vec{v}_0\, t + \vec{r}_0$, where $\vec{r}_0$ is the initial position of the particle.

Solution: We have

$$\frac{d\vec{r}}{dt} = \vec{v}_0 \tag{2.17}$$

We are given that the velocity $\vec{v}_0$ is constant. It may be possible only when the motion is along a straight line; since velocity can never be constant when the motion is along a curved path. On integrating equation (2.17), we get

$$\vec{r} = \vec{v}_0\, t + c \tag{2.18}$$

where c is the constant of integration. Using the information that we have $\vec{r} = \vec{r}_0$ when $t = 0$, we get $c = \vec{r}_0$. Using the value of c in equation (2.18), we get

$$\vec{r} = \vec{v}_0\, t + \vec{r}_0$$

2.6 Motion of a particle in a uniform gravitational field

Suppose a particle of mass m is projected from a point O on the surface of earth with coordinates (0, 0) with velocity v_0 in a direction making an angle θ with the horizontal (Figure 2.2). The magnitude of velocity is constant. The motion of a projectile is in a plane, let it be the xy-plane. Suppose, at some time t, the particle is at a point P with coordinates (x, y). The force acting on the particle is $\vec{F} = -mg\,\hat{j}$ due to the gravitational field, which is uniform having a constant value for g. Thus, the equation of motion is

$$\vec{F} = m\left[\frac{d^2x}{dt^2}\,\hat{\imath} + \frac{d^2y}{dt^2}\,\hat{\jmath}\right] \quad \text{or} \quad -mg\hat{\jmath} = m\left[\frac{d^2x}{dt^2}\,\hat{\imath} + \frac{d^2y}{dt^2}\,\hat{\jmath}\right]$$

On comparing the coefficients of $\hat{\imath}$ and $\hat{\jmath}$ on both sides of this equation, we have

$$\frac{d^2y}{dt^2} = -g \qquad \text{and} \qquad \frac{d^2x}{dt^2} = 0$$

On integration of these equations, we get

$$\frac{dy}{dt} = -gt + c_1 \qquad \text{and} \qquad \frac{dx}{dt} = c_2 \tag{2.19}$$

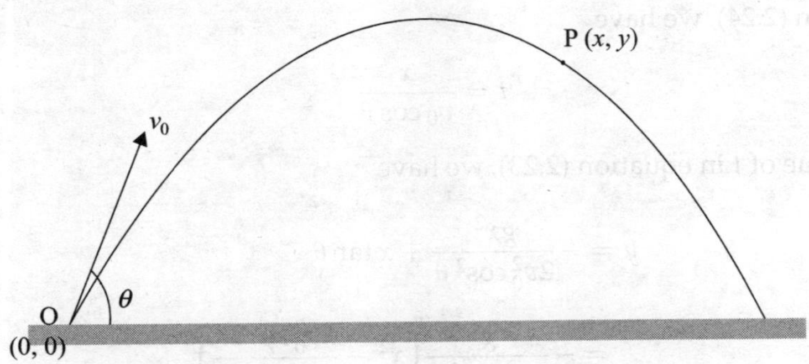

Figure 2.2: Motion of a particle under uniform gravitational field.

where c_1 and c_2 are constants of integration. The initial velocities are

$$\left(\frac{dy}{dt}\right)_{t=0} = v_0 \sin\theta \qquad \text{and} \qquad \left(\frac{dx}{dt}\right)_{t=0} = v_0 \cos\theta \tag{2.20}$$

Using equations (2.20) in (2.19), we get

$$c_1 = v_0 \sin\theta \qquad \text{and} \qquad c_2 = v_0 \cos\theta$$

Using the values of c_1 and c_2 in equations (2.19), we have

$$\frac{dy}{dt} = -gt + v_0 \sin\theta \qquad \text{and} \qquad \frac{dx}{dt} = v_0 \cos\theta$$

On integration of these equations, we get

$$y = -\frac{1}{2} gt^2 + v_0 t \sin\theta + c_3 \qquad \text{and} \qquad x = v_0 t \cos\theta + c_4 \tag{2.21}$$

where c_3 and c_4 are constants of integration. The initial positions are

$$(y)_{t=0} = 0 \qquad \text{and} \qquad (x)_{t=0} = 0 \tag{2.22}$$

Using equations (2.22) in (2.21), we get

$$0 = c_3 \qquad \text{and} \qquad 0 = c_4$$

Using the values of c_3 and c_4 in equations (2.21), we have

$$y = -\frac{1}{2} gt^2 + v_0 t \sin\theta \tag{2.23}$$

and

$$x = v_0 t \cos\theta \tag{2.24}$$

Equation (2.24) shows that the displacement along x-axis is proportional to time t.

2.6.1 Equation of orbit

From equation (2.24), we have

$$t = \frac{x}{v_0 \cos\theta}$$

Using the value of t in equation (2.23), we have

$$\begin{aligned} y &= -\frac{gx^2}{2v_0^2 \cos^2\theta} + x \tan\theta \\ &= -\frac{g}{2v_0^2 \cos^2\theta}\left(x^2 - \frac{xv_0^2 \sin 2\theta}{g}\right) \end{aligned}$$

This is equation of parabola. Hence, the equation of orbit of a projectile is a parabola.

2.6.2 Time of flight

The time of flight is the time after which the projectile comes back to the surface of the earth. In order to get the time of flight, we put $y = 0$ in equation (2.23), so that

$$0 = -\frac{1}{2} gt^2 + v_0 t \sin\theta$$

This is a quadratic equation in t and has two values for t

$$t = 0 \qquad \text{and} \qquad t = \frac{2v_0 \sin\theta}{g}$$

The first value corresponds to the initial time when the projectile was launched and the second value gives the time of flight. Thus, the time of flight T is

$$T = \frac{2v_0 \sin\theta}{g}$$

The time of flight not only depends on the initial velocity but also on the angle of projection.

2.6.3 Range

The range R is the distance travelled along x-direction in the time of flight. It can be obtained by putting the time of flight in equation (2.24) as

$$R = v_0 \cos\theta \; T$$

Using the value of T, we get

$$R = v_0 \cos\theta \, \frac{2v_0 \sin\theta}{g} = \frac{v_0^2 \sin 2\theta}{g}$$

The range will be maximum when $2\theta = 90°$. Thus, the maximum range is

$$R_{\text{max}} = \frac{v_0^2}{g}$$

For the maximum range, the angle of projection needs to be 45°.

Exercise 2.4:. A body is projected in the xy-plane from a point $(x_0,\ y_0)$ with velocity v_0 making an angle with the horizontal. Find out expression for the coordinates of a point on the path.

Solution: Suppose a particle of mass m is projected from a point O with coordinates (x_0, y_0) with velocity $\vec{v}_0$ in a direction making an angle θ with the horizontal. The magnitude of velocity is constant. The motion of a projectile is in the xy-plane. Suppose, at some time t, the particle is at a point P with coordinates $(x,\ y)$. The force acting on the particle is $\vec{F} = -mg\,\hat{j}$ due to the gravitational field, which is uniform having constant value for g. Thus, the equation of motion is

$$\vec{F} = m\left[\frac{\mathrm{d}^2x}{\mathrm{d}t^2}\,\hat{i} + \frac{\mathrm{d}^2y}{\mathrm{d}t^2}\,\hat{j}\right] \qquad \text{or} \qquad -mg\hat{j} = m\left[\frac{\mathrm{d}^2x}{\mathrm{d}t^2}\,\hat{i} + \frac{\mathrm{d}^2y}{\mathrm{d}t^2}\,\hat{j}\right]$$

On comparing the coefficients of $\hat{i}$ and $\hat{j}$ on both sides of this equation, we have

$$\frac{\mathrm{d}^2y}{\mathrm{d}t^2} = -g \qquad \text{and} \qquad \frac{\mathrm{d}^2x}{\mathrm{d}t^2} = 0$$

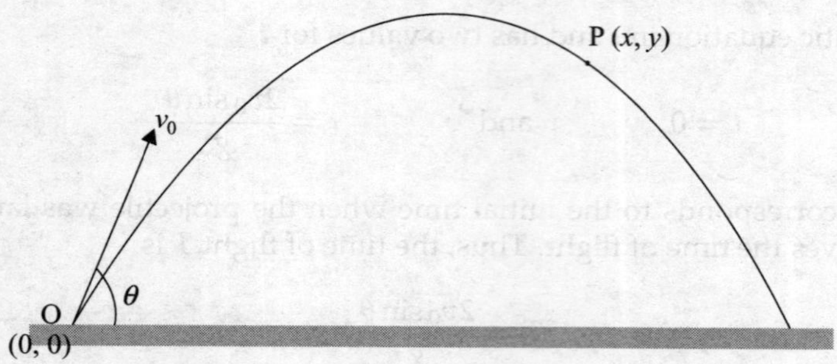

On integration of these equations, we get

$$\frac{dy}{dt} = -gt + c_1 \qquad \text{and} \qquad \frac{dx}{dt} = c_2 \tag{2.25}$$

where c_1 and c_2 are constants of integration. The initial velocities are

$$\left(\frac{dy}{dt}\right)_{t=0} = v_0 \sin\theta \qquad \text{and} \qquad \left(\frac{dx}{dt}\right)_{t=0} = v_0 \cos\theta \tag{2.26}$$

Using equations (2.26) in (2.25), we get

$$c_1 = v_0 \sin\theta \qquad \text{and} \qquad c_2 = v_0 \cos\theta$$

Using the values of c_1 and c_2 in equations (2.25), we have

$$\frac{dy}{dt} = -gt + v_0 \sin\theta \qquad \text{and} \qquad \frac{dx}{dt} = v_0 \cos\theta$$

On integration of these equations, we get

$$y = -\frac{1}{2} gt^2 + v_0 t \sin\theta + c_3 \qquad \text{and} \qquad x = v_0 t \cos\theta + c_4 \tag{2.27}$$

where c_3 and c_4 are constants of integration. The initial positions are

$$(y)_{t=0} = y_0 \qquad \text{and} \qquad (x)_{t=0} = x_0 \tag{2.28}$$

Using equations (2.28) in (2.27), we get

$$y_0 = c_3 \qquad \text{and} \qquad x_0 = c_4$$

Using the values of c_3 and c_4 in equations (2.27), we have

$$y = -\frac{1}{2} gt^2 + v_0 t \sin\theta + y_0 \tag{2.29}$$

and

$$x = v_0 t \cos\theta + x_0 \tag{2.30}$$

Equation (2.30) shows that the displacement along x-axis is proportional to time t.

Exercise 2.5:. A body is projected form the earth's surface with the velocity 20 m/s at an angle $\theta = 30°$ with the horizontal. Calculate time of flight and the range of the flight.

Solution: The time of flight is

$$T = \frac{2v_0 \sin\theta}{g} = \frac{2 \times 20 \sin 30°}{9.8} = 2.04 \text{ s}$$

The range of the flight is

$$R = \frac{v_0^2 \sin 2\theta}{g} = \frac{20^2 \sin 60°}{9.8} = 35.35 \text{ m}$$

2.7 Newton's law of gravitation

According to Newton's law of gravitation, a particle A of mass m_1 attracts any other particle B of mass m_2 with a force, called the gravitational force, expressed as

$$\vec{F} = -\frac{Gm_1m_2}{r^2}\hat{r} \tag{2.31}$$

where r is the distance between A and B, and $\hat{r}$ is the unit vector from A towards B. G is the gravitational constant, also known as a universal constant, as its value remains same in every part of the universe. Here, the sizes of A and B are much smaller than the separation between them. This is a force acting on B due to A and the minus sign indicates that the force is attractive. The force on A due to B is expressed by the same expression (2.31), but now the unit vector $\hat{r}$ is from B towards A. The force is also applicable for the heavenly bodies, like stars (sun), planets, satellites, etc. There, the separation between two bodies is much larger than the sizes of the bodies. Therefore, for applying the law of gravitation, the heavenly bodies may be treated as point masses.

The gravitational force is a central force, as it is always directed along the line joining the two point masses and depends only on the distance between the two masses.

Exercise 2.6: Suppose the mass and radius of the earth are M_e and R_e, respectively. Show that the force F acting on a particle of mass m placed at a distance r from the centre of the earth is

$$F = mg\left(\frac{R_e}{r}\right)^2$$

where g is the acceleration due to gravity of the earth.

Solution: When a particle of mass m is placed on the surface of the earth, the force on the particle is

$$mg = \frac{GmM_e}{R_e^2} \tag{2.32}$$

Here, we assume that total mass M_e of the earth is concentrated at the centre of the earth. When the particle is at a distance r from the centre of the earth, the force on the particle is

$$F = \frac{GmM_e}{r^2} \tag{2.33}$$

On dividing equation (2.33) by (2.32), we get

$$\frac{F}{mg} = \frac{R_e^2}{r^2} \qquad \text{or} \qquad F = mg\left(\frac{R_e}{r}\right)^2$$

Note: It is known experimentally to a high degree of accuracy that gravitational mass and inertial mass of a body are equal. The mass in the gravitational equation is known as the gravitational mass, whereas the mass in Newton's second law of motion is known as the inertial mass.

2.7.1 Motion of a satellite in a circular orbit

A satellite is a body that revolves around a planet. A satellite may be natural or artificial. Moon is a natural satellite revolving around the earth (planet). A large number of man-made (artificial) satellites are revolving around the earth. For simplicity, we assume that a satellite of mass m is revolving around the earth in a circular orbit of radius r. As the earth spins from west towards east, a satellite revolving around the earth also kept to move from west towards east, so that the satellite is always above its country. The magnitude of the force on the satellite due to the earth of mass M_e is

$$F = \frac{GmM_e}{r^2}$$

This force is directed towards the earth. Owing to the circular (curved) path, a force called the centrifugal force $m\omega^2 r$ acts on the satellite in the outward direction. Here, ω is the angular velocity of revolution of the satellite. These two oppositely directed forces balance each other, and we have

$$\frac{GmM_e}{r^2} = m\omega^2 r \tag{2.34}$$

The time period of revolution of the satellite is

$$T = \frac{2\pi}{\omega} \qquad \text{or} \qquad \omega = \frac{2\pi}{T} \tag{2.35}$$

Using equation (2.35) in (2.34) and rearranging, we get

$$r^3 = \frac{GM_e T^2}{4\pi^2}$$

It shows that the square of time period of revolution is proportional to the cube of the radius of the path of revolution.

Geostationary satellite

For a geostationary satellite, the time period of its revolution must be equal to the time period of rotation of the earth. That is, $T = 1$ day. Here, we adjust the distance of satellite from the earth such that the time period of its revolution is 1 day. It revolves from west towards east, so that it is always situated above the country it belongs to.

Exercise 2.7: Calculate distance of geostationary satellite from the centre of the earth. Given the mass of earth $M_e = 5.98 \times 10^{24}$ kg, time period of rotation of earth $T = 1$ day and $G = 6.67 \times 10^{-11}$ N-m^2/kg^2.

Solution: Here, we shall assume that rotation period of the earth about its own axis is 24 hr (1 day). We have

$$T = 1 \text{ day} = 1 \times 24 \times 60 \times 60 = 8.64 \times 10^4 \text{ s}$$

The distance r of the satellite from the centre of the earth is expressed as

$$r^3 = \frac{GM_e T^2}{4\pi^2} = \frac{6.67 \times 10^{-11} \times 5.98 \times 10^{24} (8.64 \times 10^4)^2}{4 \times 3.14^2}$$

$$= 7.50 \times 10^{22}$$

or

$$r = (7.50 \times 10^{22})^{1/3} = 4.2 \times 10^7 \text{ m}$$

Thus, a geostationary satellite will be at a distance of 4.2×10^7 m from the centre of the earth. The radius of the earth is 6.4×10^6 m. So the altitude (distance from the earth's surface) of the geostationary satellite is $h = 4.2 \times 10^7 - 6.4 \times 10^6 = 3.56 \times 10^7$ m.

2.8 Charged particle in electric and magnetic fields

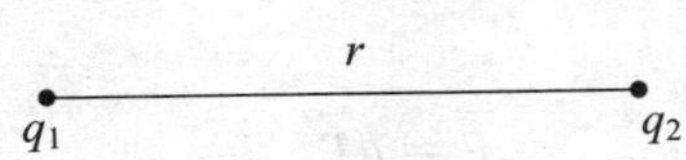

Figure 2.3: Two point charges q_1 and q_2 separated by a distance r.

Suppose two particles A and B carrying charges q_1 and q_2, respectively, are separated by a distance r (Figure 2.3), then the electrostatic force acting on B due to A is

$$\vec{F} = \frac{1}{4\pi\epsilon_0} \frac{q_1 q_2}{r^2} \hat{r} \tag{2.36}$$

where r is the distance between A and B, and the unit vector $\hat{r}$ is from A towards B. The sizes of A and B are much smaller than the separation between them. The force is repulsive when both the charges have the same sign and attractive when the charges have opposite sign. Further, the force acting on A due to B is given by the same expression (Equation 2.36) where the unit vector $\hat{r}$ is from B towards A.

Exercise 2.8: Calculate the force between two protons separated by a distance 10^{-14} m. Given $(4\pi\epsilon_0)^{-1} = 8.99 \times 10^9$ N-m^2/C^2, charge of proton $= 1.60 \times 10^{-19}$ C.

Solution: If q is the charge on a proton, the force between two protons is

$$F = \frac{1}{4\pi\epsilon_0} \frac{qq}{r^2} = \frac{8.99 \times 10^9 (1.60 \times 10^{-19})^2}{(10^{-14})^2} = 2.3 \text{ N}$$

2.8.1 Electric field

Suppose a particle carrying a charge q is situated in an external electric field $\vec{E}$. The electric field intensity $\vec{E}$ and the force $\vec{F}$ on the particle carrying charge q and placed in the field is expressed as

$$\vec{F} = q\vec{E} \tag{2.37}$$

For two charges q_1 and q_2 separated by a distance r, we have

$$\vec{F} = \frac{1}{4\pi\epsilon_0}\frac{q_1 q_2}{r^2}\hat{r} \tag{2.38}$$

Suppose q is one of the charges, say $q = q_2$, then from equations (2.38) and (2.37), we have

$$q_2\,\vec{E} = \frac{1}{4\pi\epsilon_0}\frac{q_1 q_2}{r^2}\hat{r} \qquad \text{or} \qquad \vec{E} = \frac{1}{4\pi\epsilon_0}\frac{q_1}{r^2}\hat{r}$$

This is electric field $\vec{E}$ due to a charge q_1 at distance r. Here, $\hat{r}$ is the unit vector in the direction away from the charge q_1.

2.8.2 Motion of a charged particle in a uniform electric field

Suppose a particle of mass m and carrying a charge q is placed in an electric field $\vec{E}$. Then we have

$$\vec{F} = q\vec{E} \tag{2.39}$$

Now, the force can be expressed as

$$\vec{F} = m\vec{a} \tag{2.40}$$

where $\vec{a}$ is acceleration and can be expressed as

$$\vec{a} = \frac{d^2\vec{r}}{dt^2} \tag{2.41}$$

where $\vec{r}$ is position vector of the particle relative to the origin. From equations (2.39), (2.40), (2.41), we have

$$\frac{d^2\vec{r}}{dt^2} = \frac{\vec{F}}{m} \qquad \text{or} \qquad \frac{d^2\vec{r}}{dt^2} = \frac{q\vec{E}}{m}$$

As the electric field is uniform, on integration we have

$$\frac{d\vec{r}}{dt} = \frac{q\vec{E}}{m}t + c_1 \tag{2.42}$$

where c_1 is the constant of integration. On applying the condition that at $t = 0$, the velocity is $\vec{v}_0$, we get $c_1 = \vec{v}_0$. Using this value in equation (2.42), we get

$$\frac{d\vec{r}}{dt} = \frac{q\vec{E}}{m}t + \vec{v}_0$$

On integration, we have

$$\vec{r} = \frac{q\vec{E}}{2m}t^2 + \vec{v}_0 t + c_2 \tag{2.43}$$

where c_2 is the constant of integration. On applying the condition that at $t = 0$, the displacement is $\vec{r}_0$, we get $c_2 = \vec{r}_0$. Using this value in equation (2.43), we get

$$\vec{r} = \frac{q\vec{E}}{2m}t^2 + \vec{v}_0 t + \vec{r}_0$$

This equation gives the position vector $\vec{r}$ at time t of a charged particle moving in a uniform electric field $\vec{E}$. The position of the particle does not change uniformly with time, as the first term on right side is proportional to t^2. When $\vec{v}_0 = 0$ and $\vec{r}_0 = 0$, we have $\vec{r}$ along $\vec{E}$.

2.8.3 Motion of a charged particle in a uniform magnetic field

Suppose a particle of mass m, carrying a charge q and moving with velocity $\vec{v}$ is placed in a magnetic field $\vec{B}$. The force $\vec{F}$ acting on the particle is

$$\vec{F} = q\vec{v} \times \vec{B} \tag{2.44}$$

Now, the force can be expressed as

$$\vec{F} = m\vec{a} = m\frac{d\vec{v}}{dt} \tag{2.45}$$

where $\vec{a}$ is the acceleration of the particle. From equations (2.44) and (2.45), the equation of motion is

$$m\frac{d\vec{v}}{dt} = q\vec{v} \times \vec{B} \tag{2.46}$$

Let the uniform magnetic field be directed along the z-axis, so that

$$\vec{B} = B\hat{k}$$

For the velocity expressed as $\vec{v} = v_x\hat{i} + v_y\hat{j} + v_z\hat{k}$, we have

$$\vec{v} \times \vec{B} = \begin{vmatrix} \hat{i} & \hat{j} & \hat{k} \\ v_x & v_y & v_z \\ 0 & 0 & B \end{vmatrix} = v_yB\hat{i} - v_xB\hat{j}$$

Thus, equation (2.46) can be expressed as

$$m\,\frac{\mathrm{d}}{\mathrm{d}t}(v_x\hat{i}+v_y\hat{j}+v_z\hat{k}) = q(v_yB\hat{i}-v_xB\hat{j})$$

On comparing the coefficients of $\hat{i}$, $\hat{j}$, $\hat{k}$ on the two sides of this equation, we get

$$\frac{dv_x}{dt} = \frac{qv_yB}{m} \qquad \frac{dv_y}{dt} = -\frac{qv_xB}{m} \qquad \frac{dv_z}{dt} = 0 \tag{2.47}$$

Form the relation

$$\frac{dv_z}{dt} = 0$$

we get v_z = constant, showing that component of velocity along the z-axis is constant.

Kinetic energy

The kinetic energy is expressed as

$$K = \frac{1}{2}\,m\vec{v}\,.\,\vec{v}$$

The rate of change of kinetic energy is

$$\frac{dK}{dt} = \frac{1}{2}m\vec{v}\,.\frac{d\vec{v}}{dt} + \frac{1}{2}\,m\frac{d\vec{v}}{dt}.\,\vec{v} = m\vec{v}\,.\frac{d\vec{v}}{dt}$$

Using equation (2.46) here, we get

$$\frac{dK}{dt} = \vec{v}\,.\,q(\vec{v}\times\vec{B}) = 0$$

The value obtained is zero because the direction of $(\vec{v}\times\vec{B})$ is perpendicular to the direction of $\vec{v}$ and the dot product of $\vec{v}$ and $(\vec{v}\times\vec{B})$ will give zero. It shows that the kinetic energy of a charged particle moving in a uniform magnetic field is conserved. The magnetic field therefore does no work; it only changes the direction of motion of the particle.

Orbit of motion

Equation (2.47) shows that derivative of v_x is proportional to v_y and derivative of v_y is proportional to v_x. Such situation is found in case of sine and cosine functions. Let us look for the solutions

$$v_x = v_1\sin\omega t \qquad v_y = v_1\cos\omega t \qquad v_z = \text{constant} \tag{2.48}$$

Using equation (2.48) in (2.47), we get

$$v_1\omega\cos\omega t = \frac{qB}{m}\,v_1\cos\omega t \qquad -v_1\omega\sin\omega t = -\frac{qB}{m}\,v_1\sin\omega t$$

this gives

$$\omega = \frac{qB}{m} = \omega_c$$

where $\omega_c = qB/m$ is known as the cyclotron frequency. Using the value of ω in equation (2.48), we have

$$v_x = v_1 \sin \omega_c t \qquad v_y = v_1 \cos \omega_c t \qquad v_z = \text{constant} \tag{2.49}$$

If we assume that the charged particle is moving along y-axis, so that $v_y(0) = v_0$, we have

$$v_x = 0 \qquad v_0 = v_1 \qquad v_z = 0$$

and equation (2.49) modifies to

$$v_x = v_0 \sin \omega_c t \qquad v_y = v_0 \cos \omega_c t \qquad v_z = 0 \tag{2.50}$$

Integration of equation (2.50) gives

$$x = -\frac{v_0}{\omega_c} \cos \omega_c t + c_1 \qquad y = \frac{v_0}{\omega_c} \sin \omega_c t + c_2 \qquad z = c_3 \tag{2.51}$$

where c_1, c_2, c_3 are constants of integration. Suppose at $t = 0$ we have $x = x_0$, $y = y_0$, $z = z_0$. From equation (2.51), we have

$$x_0 = -\frac{v_0}{\omega_c} + c_1 \qquad y_0 = c_2 \qquad z_0 = c_3$$

Using the values of c_1, c_2, c_3 in equation (2.51), we have

$$x = -\frac{v_0}{\omega_c} \cos \omega_c t + x_0 + \frac{v_0}{\omega_c} \qquad y = \frac{v_0}{\omega_c} \sin \omega_c t + y_0 \qquad z = z_0$$

It gives

$$(x - x_0 - \frac{v_0}{\omega_c})^2 + (y - y_0)^2 = \frac{v_0^2}{\omega_c^2}$$

Showing that the orbit of motion is a circle whose centre is at $(x_0 + v_0/c, y_0, z_0)$.

Exercise 2.9: Calculate the cyclotron frequency of an electron and of a proton moving in a magnetic field of 1.2 Tesla. Given the mass of electron $m_e = 9.1 \times 10^{-31}$ kg, charge of electron $e = 1.6 \times 10^{-19}$ C and mass of proton $m_p = 1.6 \times 10^{-27}$ kg.

Solution: The cyclotron frequency of an electron is

$$\omega_c = \frac{eB}{m_e} = \frac{1.6 \times 10^{-19} \times 1.2}{9.1 \times 10^{-31}} = 2.1 \times 10^{11}\ \text{s}^{-1}$$

The cyclotron frequency of proton is

$$\omega_c = \frac{eB}{m_p} = \frac{1.6 \times 10^{-19} \times 1.2}{1.6 \times 10^{-27}} = 1.2 \times 10^{8}\ \text{s}^{-1}$$

2.9 Potential energy

For a force $\vec{F}$, the corresponding potential energy V is expressed as

$$\vec{F} = -\nabla V$$

For one-dimensional case, we have

$$\vec{F} = -\frac{\partial V}{\partial r}\,\hat{r}$$

When we account only for magnitude, we have

$$F = -\frac{\partial V}{\partial r}$$

Exercise 2.10:. Derive an expression for the potential energy corresponding to the gravitational force.

Solution: The gravitational force F between two bodies of masses m_1 and m_2 separated by a distance r is

$$F = -\frac{Gm_1m_2}{r^2}$$

Here, the minus sign indicates that the force is attractive. Since the gravitational force is along the line joining the two masses, the conservative force F and the potential energy V are related through the expression

$$F = -\frac{dV}{dr}$$

Therefore,

$$-\frac{Gm_1m_2}{r^2} = -\frac{dV}{dr} \quad \text{or} \quad dV = \frac{Gm_1m_2}{r^2}\,dr$$

On integration, we get

$$V = -\frac{Gm_1m_2}{r} + c$$

where c is the constant of integration. Conventionally, we take $V = 0$, at $r = \infty$. It gives $c = 0$, and the gravitational potential energy of a system of two masses m_1 and m_2 separated by a distance r is

$$V = -\frac{Gm_1m_2}{r}$$

Exercise 2.11:. Derive an expression for the potential energy corresponding to the Coulomb force.

Solution: The Coulomb force between two charges q_1 and q_2 separated by a distance r is

$$F = \frac{1}{4\pi\epsilon_0}\frac{q_1 q_2}{r^2}$$

The force is positive (repulsive) when the charges have the same sign and negative (attractive) when the charges have opposite sign. Since the Coulomb force is along the line joining the two charges, the conservative force F and the potential energy V are related through the expression

$$F = -\frac{dV}{dr}$$

Therefore,

$$\frac{1}{4\pi\epsilon_0}\frac{q_1 q_2}{r^2} = -\frac{dV}{dr} \quad \text{or} \quad dV = -\frac{1}{4\pi\epsilon_0}\frac{q_1 q_2}{r^2}\, dr$$

On integration, we get

$$V = \frac{1}{4\pi\epsilon_0}\frac{q_1 q_2}{r} + c$$

where c is the constant of integration. Conventionally, we take $V = 0$, at $r = \infty$. It gives $c = 0$, and the Coulomb potential energy of a system of two charges q_1 and q_2 separated by a distance r is

$$V = \frac{1}{4\pi\epsilon_0}\frac{q_1 q_2}{r}$$

The potential energy is positive for a repulsive force and negative for an attractive force.

Multiple-choice questions

1. For Galilean transformations, which of the following statements is wrong?

(i) Time and space are not coupled. (ii) Time is absolute.

(iii) Acceleration is invariant. (iv) Velocity is invariant.

Ans. (iv)

2. Which of the following relations is the correct expression for Newton's second law of motion?

(i) $\vec{F} = d\vec{p}/dt$ (ii) $\vec{F} = m\vec{a}/2$

(iii) $\vec{F} = q\,\vec{v} \times \vec{B}$ (iv) $E = mv^2/2$

Ans. (i)

3. For a projectile, the range is maximum when the projection angle is

(i) 0° (ii) 30°

(iii) 45° (iv) 60°

Ans. (iii)

4. For a projectile with initial velocity v_0 and projected at an angle θ, the time of flight T is

   (i) $T = 2v_0 \sin\theta/g$ (ii) $T = v_0/g$

   (iii) $T = v_0^2 \sin\theta/g$ (iv) $T = v_0 \sin\theta$

   Ans. (i)

5. For a projectile with initial velocity v_0 and projected at an angle θ, the range R is

   (i) $R = v_0^2/2g$ (ii) $R = v_0^2 \sin\theta/g$

   (iii) $R = v_0^2 \sin 2\theta/g$ (iv) $R = v_0 \sin 2\theta/g$

   Ans. (iii)

6. Which one of the following is non-conservative force?

   (i) Gravitational force (ii) Electrostatic force

   (iii) Lorentz force (iv) Magnetic force

   Ans. (iii)

7. Which one of the following is a conservative force?

   (i) Lorentz force (ii) Frictional force

   (iii) Coriolis force (iv) Gravitational force

   Ans. (iv)

Exercises

1. Describe inertial and non-inertial reference frames.
2. State Newton's laws of motion.
3. Define gravitational mass and inertial mass.
4. Derive law of conservation of momentum using Newton's third law of motion.
5. A particle of mass m is projected from a point on the surface of earth with velocity $\vec{v}_0$ in a direction making an angle θ with the horizontal. Derive expressions for the time of flight, range, maximum range and for the path of the motion.
6. For a body projected from the earth's surface, show that the range is maximum when the angle of projection is 45°.
7. Show that for a body projected from the earth's surface the time of flight and range do not depend on the mass of the projectile.
8. Define gravitational mass and inertial mass of a particle. How they are related?
9. Describe motion of a charged particle in a uniform constant electric field.

10. Show that for a charged particle moving in a uniform magnetic field, the kinetic energy of the particle does not change.

11. Show that for a charged particle moving in a uniform magnetic field, the orbit of motion is a circle.

12. Describe motion of a satellite revolving around the earth in a circular orbit.

13. What is a geostationary satellite? Calculate the altitude of geostationary satellite.

14. Show that for a particle moving along a circular (curved) path, the force acting on it can never be zero.

15. What is an inertial reference frame? Describe the reason that a reference frame attached to the earth's surface may be treated as an inertial reference frame.

16. Describe inertial and non-inertial reference frames.

17. What are conservative and non-conservative forces? Give examples of each of them.

18. What are Galilean transformations? Show that under Galilean transformations the acceleration of a particle is invariant.

19. Discuss conservative and non-conservative forces.

20. Write notes on the following:

    (i) Laboratory system of reference
    (ii) Centre-of-mass system
    (iii) Newton's laws of motion
    (iv) Newton's first law of motion
    (v) Newton's second law of motion
    (vi) Newton's third law of motion
    (vii) Time of flight and range of a projectile
    (viii) Newton's law of universal gravitation
    (ix) Motion of a satellite in a circular orbit
    (x) Motion of a charged particle in a uniform magnetic field
    (xi) Geostationary satellite
    (xii) Inertial reference frame
    (xiii) Conservative and non-conservative forces

CHAPTER 3

CONSERVATION LAWS

A conservation law in the physical world is a consequence of some symmetry. A number of conservation laws exist. Some of them are exact and some are approximate. There are conservation laws pertaining to energy, momentum, angular momentum, charge, number of baryons (protons, neutrons and heavier elements), strangeness and various quantities. In this book, we are mainly interested in the conservation of energy, momentum and angular momentum. The conservation laws are powerful tools because of the following:

1. Conservation laws are independent of the details of trajectory and often of the details of the particular force.

2. Conservation laws may be used even when the force is not known. This applies particularly in the physics of elementary particles.

3. Conservation laws have an intimate connection with invariance.

4. Even when the force is known exactly, a conservation law may be a convenient help in solving for the motion of a particle.

3.1 Elastic and inelastic collisions

Generally, we consider collision between two bodies. A collision between two bodies may be either elastic or inelastic. In an elastic collision, total kinetic energy of the two bodies before collision is equal to the total kinetic energy of the bodies after collision. That is, in an elastic collision, the kinetic energy of the system of bodies is conserved. However, the kinetic energy may be shared among the bodies during collision. On the other hand, in an inelastic collision, the kinetic energy of the system is not conserved. A part of kinetic energy, after collision, is converted to some other forms of energy or other forms of energy may be converted into the kinetic energy. For example, internal energy of some bodies may change (body may be excited to a higher energy state or deexcited to a lower energy state). It is worth mentioning that in both the elastic and inelastic collisions, total mass and total momentum of the system of two bodies remain conserved. However, total kinetic energy is conserved for the elastic collision, and not conserved for the inelastic collision.

3.1.1 Elastic collision

Let us consider two bodies 1 and 2 of masses m_1 and m_2. Suppose body 1 is moving with velocity v_1 and body 2 is stationary (for convenience). We can consider a coordinate system so that x-axis is along the direction of motion of body 1. Suppose after collision, body 1 is moving with velocity v_1' in the direction θ and body 2 is moving with velocity v_2' in the direction ϕ, as shown in Figure 3.1. For elastic collision, total mass, total momentum and total kinetic energy of the system are conserved.

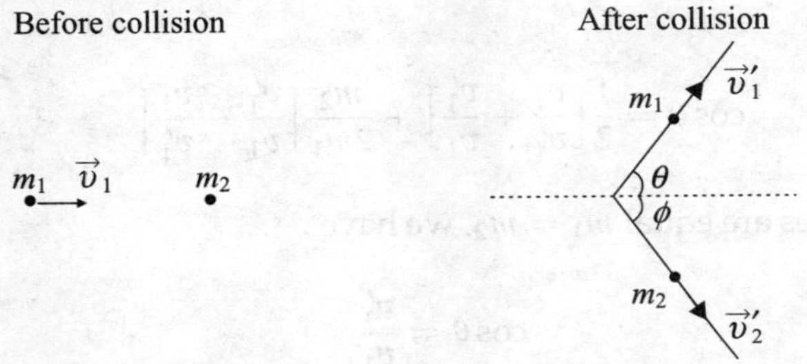

Figure 3.1: Collision of two bodies in the laboratory system of reference.

For the elastic collision, from the conservation of kinetic energy we have

$$\frac{1}{2}\, m_1 v_1^2 = \frac{1}{2}\, m_1 v_1'^2 + \frac{1}{2}\, m_2 v_2'^2 \tag{3.1}$$

The masses m_1 and m_2 of the two bodies are not changing. Further, the total momentum is conserved. Since momentum is a vector quantity, we resolve it into two components, one along the x-axis and the other along the y-axis. From the conservation of momentum along x-axis, we have

$$m_1 v_1 = m_1 v_1' \cos\theta + m_2 v_2' \cos\phi \tag{3.2}$$

and along y-axis, we have

$$0 = m_1 v_1' \sin\theta - m_2 v_2' \sin\phi \tag{3.3}$$

From equations (3.2) and (3.3), we have

$$m_2 v_2' \cos\phi = m_1 v_1 - m_1 v_1' \cos\theta \tag{3.4}$$

and

$$m_2 v_2' \sin\phi = m_1 v_1' \sin\theta \tag{3.5}$$

On squaring equations (3.4) and (3.5), and adding the results, we get

$$m_2^2 v_2'^2 = m_1^2 v_1^2 + m_1^2 v_1'^2 - 2m_1^2 v_1 v_1' \cos\theta \tag{3.6}$$

On multiplying equation (3.1) by $2m_2$ and rearranging, we get

$$m_2^2 v_2'^2 = m_1 m_2 v_1^2 - m_1 m_2 v_1'^2 \tag{3.7}$$

Left sides of equations (3.6) and (3.7) are equal and therefore, we have

$$m_1 m_2 v_1^2 - m_1 m_2 v_1'^2 = m_1^2 v_1^2 + m_1^2 v_1'^2 - 2m_1^2 v_1 v_1' \cos\theta$$

On rearranging this equation, we get

$$\cos\theta = \frac{m_1^2(v_1^2 + v_1'^2) + m_1 m_2(v_1'^2 - v_1^2)}{2m_1^2 v_1 v_1'}$$

or

$$\cos\theta = \frac{1}{2}\left[\frac{v_1}{v_1'} + \frac{v_1'}{v_1}\right] + \frac{m_2}{2m_1}\left[\frac{v_1'}{v_1} - \frac{v_1}{v_1'}\right]$$

When the two masses are equal $m_1 = m_2$, we have

$$\cos\theta = \frac{v_1'}{v_1}$$

For the case $m_1 = m_2$, since $0 < v_1' < v_1$, the value of angle θ can take the values $0 < \theta < 90°$.

3.1.2 Collision in centre-of-mass system

Let us now look at this scattering process in the centre-of-mass system. The velocity of the centre of mass is

$$\vec{V} = \frac{m_1 \vec{v_1}}{m_1 + m_2} \tag{3.8}$$

In the centre-of-mass system, the velocities of bodies before collision are

$$\vec{u}_1 = \vec{v}_1 - \vec{V} \quad \text{and} \quad \vec{u}_2 = \vec{v}_2 - \vec{V} \tag{3.9}$$

We know that in the centre-of-mass system, the momenta of two particles are opposite to each other, as shown in Figure 3.2. Thus, the total momentum of the system in the centre-of-mass system is zero.

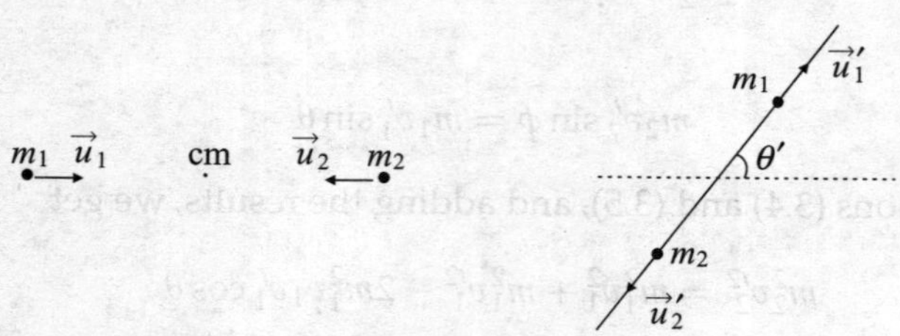

Figure 3.2: Elastic collision in the centre-of-mass system.
Note: cm denotes the centre of mass of the system.

In the centre-of-mass system, since velocities are measured with respect to the centre of mass, we have $\vec{V} = \vec{V}'$, where $\vec{V}'$ is the velocity of the centre of mass after collision. After collision, the velocities of the particles $\vec{u}_1'$ and $\vec{u}_2'$ are

$$\vec{u}_1' = \vec{v}_1' - \vec{V} \qquad \text{and} \qquad \vec{u}_2' = \vec{v}_2' - \vec{V}$$

Following the law of conservation of momentum, total momentum of the system at the centre of mass is zero. Thus, after collision, the two particles will move in the directions which are opposite to each other. In the centre-of-mass system, from the conservation of kinetic energy we have

$$\frac{1}{2}\,m_1 u_1^2 + \frac{1}{2}\,m_2 u_2^2 = \frac{1}{2}\,m_1 u_1'^2 + \frac{1}{2}\,m_2 u_2'^2$$

This equation shows that $u_1 = u_1'$ and $u_2 = u_2'$. That is, for elastic scattering in the centre of mass system, the magnitudes of velocities of the particles remain unchanged and only their directions change.

Figure 3.3: Resolution of motion of body 1 after collision in the laboratory system of reference (a) and in the centre-of-mass system (b).

On resolving the relation $\vec{v}_1' = \vec{u}_1 + \vec{V}$ into two components, we have (Figure 3.3)

$$v_1' \cos\theta = u_1 \cos\theta' + V \qquad \text{and} \qquad v_1' \sin\theta = u_1 \sin\theta'$$

From these relations, we get

$$\tan\theta = \frac{u_1 \sin\theta'}{u_1 \cos\theta' + V} \qquad \text{or} \qquad \tan\theta = \frac{\sin\theta'}{\cos\theta' + V/u_1} \tag{3.10}$$

Now, from equations (3.8) and (3.9), we have

$$\vec{V} = \frac{m_1(\vec{u}_1 + \vec{V})}{m_1 + m_2} \qquad \text{or} \qquad \vec{V} = \frac{m_1 \vec{u}_1}{m_1 + m_2} + \frac{m_1 \vec{V}}{m_1 + m_2}$$

It can be simplified to get

$$\vec{V} = \frac{m_1}{m_2}\,\vec{u}_1 \qquad \text{or} \qquad V = \frac{m_1}{m_2}\,u_1 \tag{3.11}$$

as both $\vec{V}$ and $\vec{u}_1$ are along the same direction. Using equation (3.11) in (3.10), we get

$$\tan\theta = \frac{\sin\theta'}{\cos\theta' + m_1/m_2}$$

Case 1: When the masses of two bodies are equal, $m_1 = m_2$, we have

$$\tan\theta = \frac{\sin\theta'}{\cos\theta' + 1} = \frac{2\sin\theta'/2\cos\theta'/2}{2\cos^2\theta'/2 - 1 + 1} = \tan\theta'/2$$

Hence, $\theta = \theta'/2$. It shows that for two bodies of equal masses, the scattering angle in the laboratory system of reference is half of that in the centre-of-mass system.

Case 2: When the mass of m_1 is much smaller than m_2, $m_1 << m_2$, we have

$$\tan\theta = \frac{\sin\theta'}{\cos\theta' + 0} = \tan\theta'$$

Hence, $\theta = \theta'$. It shows that when the mass of projectile is much smaller than that of target, the scattering angle in the laboratory system of reference is equal to that in the centre-of-mass system.

Exercise 3.1: Particle A of mass m moving with velocity v makes an elastic collision with a stationary particle B of equal mass m. After collision, the particle A moves with velocity v_1 in the direction making an angle θ with the initial direction and particle B moves with velocity v_2 in the direction making an angle ϕ with the initial direction in another direction. Show that

$$v_1 = v\cos\theta \qquad \text{and} \qquad v_2 = \frac{v\sin\theta\,\cos\theta}{\sin\phi}$$

Solution: Suppose after collision, the particle A is moving in the direction making an angle θ with the initial direction and the particle B is moving in the direction making an angle ϕ with the initial direction.

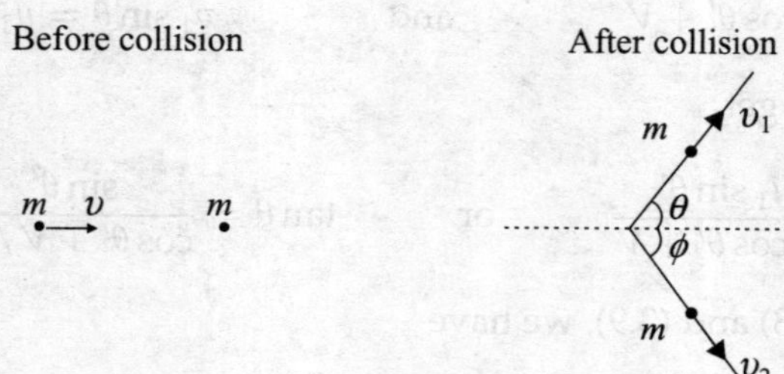

For the elastic collision, from the conservation of kinetic energy we have

$$\frac{1}{2}mv^2 = \frac{1}{2}mv_1^2 + \frac{1}{2}mv_2^2 \qquad \text{giving} \qquad v_2^2 = v^2 - v_1^2 \tag{3.12}$$

From the conservation of momentum along x-axis, we have

$$mv = mv_1\cos\theta + mv_2\cos\phi \quad \text{giving} \quad v_2\cos\phi = v - v_1\cos\theta \tag{3.13}$$

From the conservation of momentum along y-axis, we have

$$0 = mv_1 \sin\theta - mv_2 \sin\phi \qquad \text{giving} \qquad v_2 \sin\phi = v_1 \sin\theta \tag{3.14}$$

On squaring equations (3.13) and (3.14) and adding, we get

$$v_2^2 = v^2 + v_1^2 - 2vv_1 \cos\theta \tag{3.15}$$

From equations (3.12) and (3.15), we get

$$v^2 - v_1^2 = v^2 + v_1^2 - 2vv_1 \cos\theta \qquad \text{or} \qquad 2v_1^2 = 2vv_1 \cos\theta$$

It gives

$$v_1 = v\cos\theta \tag{3.16}$$

Using equation (3.16) in (3.14), we get

$$v_2 \sin\phi = v\cos\theta \sin\theta \qquad \text{or} \qquad v_2 = \frac{v \sin\theta\, \cos\theta}{\sin\phi}$$

Exercise 3.2: Particle A of mass m moving with velocity v makes an elastic collision with a stationary particle B of equal mass m. After collision, if both particles are moving with equal velocity v_1, show that each of the particles will move in a direction making an angle $\pi/4$ with the initial direction.

Solution: Suppose after collision, one particle is moving in the direction making an angle θ with the initial direction and other particle is moving in the direction making an angle ϕ with the initial direction.

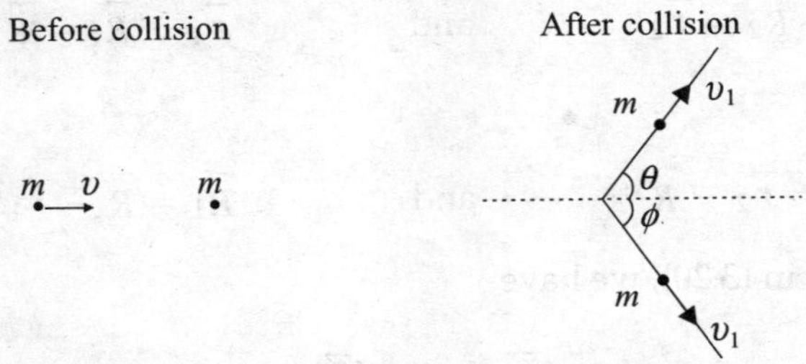

For elastic collision, from the conservation of kinetic energy we have

$$\frac{1}{2}\,mv^2 = \frac{1}{2}\,mv_1^2 + \frac{1}{2}\,mv_1^2 \qquad \text{giving} \qquad v = v_1\sqrt{2} \tag{3.17}$$

From the conservation of momentum along x-axis, we have

$$mv = mv_1 \cos\theta + mv_1 \cos\phi \qquad \text{giving} \qquad v = v_1(\cos\theta + \cos\phi) \tag{3.18}$$

From the conservation of momentum along y-axis, we have

$$0 = mv_1 \sin\theta - mv_1 \sin\phi \qquad \text{giving} \qquad \phi = \theta \tag{3.19}$$

Using equations (3.17) and (3.19) in (3.18), we get

$$v_1\sqrt{2} = v_1(\cos\theta + \cos\theta) \qquad \text{or} \qquad \cos\theta = 1/\sqrt{2}$$

Thus, we have

$$\theta = \pi/4 \qquad \text{and} \qquad \phi = \pi/4$$

3.2 Centre of mass

Let us consider two bodies 1 and 2 of masses m_1 and m_2 placed at two points A and B whose position vectors are $\vec{r}_1$ and $\vec{r}_2$, respectively. The centre of masses is at a point C in between the points A and B, as shown in Figure 3.4.

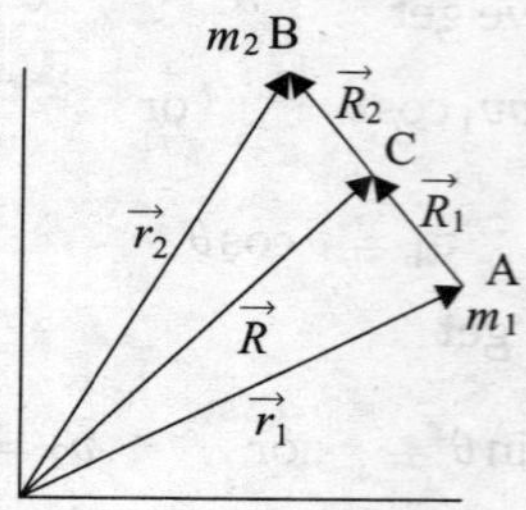

Figure 3.4: Centre of masses m_1 and m_2 is at point C.

Form the principle of balance, we have

$$m_1 R_1 = m_2 R_2 \qquad \text{or} \qquad m_1 \vec{R}_1 = m_2 \vec{R}_2 \tag{3.20}$$

as the vectors $\vec{R}_1$ and $\vec{R}_2$ are along the same direction. Now, from Figure 3.4, we have

$$\vec{R} + \vec{R}_2 = \vec{r}_2 \qquad \text{and} \qquad \vec{r}_1 + \vec{R}_1 = \vec{R}$$

Therefore,

$$\vec{R}_2 = \vec{r}_2 - \vec{R} \qquad \text{and} \qquad \vec{R}_1 = \vec{R} - \vec{r}_1 \tag{3.21}$$

Using equation (3.21) in (3.20), we have

$$m_1(\vec{R} - \vec{r}_1) = m_2(\vec{r}_2 - \vec{R})$$

On rearranging this equation, we get

$$(m_1 + m_2)\,\vec{R} = m_1 \vec{r}_1 + m_2 \vec{r}_2 \qquad \text{or} \qquad \vec{R} = \frac{m_1 \vec{r}_1 + m_2 \vec{r}_2}{m_1 + m_2}$$

This is the position vector of the centre of masses m_1 and m_2. Thus, a mass $(m_1 + m_2)$ may be considered as situated at point C whose position vector is $\vec{R}$.

Suppose, we have three bodies 1, 2, 3 of masses m_1, m_2, m_3 situated at the positions $\vec{r}_1$, $\vec{r}_2$, $\vec{r}_3$. For two bodies 1 and 2 of masses m_1, m_2 situated at the positions $\vec{r}_1$, $\vec{r}_2$, we have seen that it is equivalent to a mass $(m_1 + m_2)$ situated at the position $\vec{R}'$ expressed as

$$\vec{R}' = \frac{m_1 \vec{r}_1 + m_2 \vec{r}_2}{m_1 + m_2} \tag{3.22}$$

Hence, the given system of three bodies is equivalent to a system of two bodies of masses $(m_1 + m_2)$ and m_3 situated at the positions $\vec{R}_1$ and $\vec{r}_3$. Thus, the position vector of these two bodies is

$$\vec{R} = \frac{(m_1 + m_2)\ \vec{R}' + m_3 \vec{r}_3}{(m_1 + m_2) + m_3} \tag{3.23}$$

Using equation (3.22) in (3.23), we get

$$\vec{R} = \frac{(m_1 + m_2)\frac{m_1\vec{r}_1 + m_2\vec{r}_2}{m_1 + m_2} + m_3 \vec{r}_3}{(m_1 + m_2) + m_3}$$

$$= \frac{m_1 \vec{r}_1 + m_2 \vec{r}_2 + m_3 \vec{r}_3}{m_1 + m_2 + m_3}$$

Similarly, we can go on adding various masses situated at different positions. Finally, when n bodies of masses $m_1, m_2, \ldots, m_n$ are placed at the positions $\vec{r}_1, \vec{r}_2, \ldots, \vec{r}_n$, respectively. The position of the centre of mass is

$$\vec{R} = \frac{m_1 \vec{r}_1 + m_2 \vec{r}_2 + \ldots + m_n \vec{r}_n}{m_1 + m_2 + \ldots + m_n} \tag{3.24}$$

Exercise 3.3: Suppose three masses 1, 2 and 3 kg are situated at the positions given by $\vec{r}_1 = 2\hat{i} + \hat{j} - 4\hat{k}$, $\vec{r}_2 = \hat{i} + 2\hat{j} - \hat{k}$ and $\vec{r}_3 = 3\hat{i} + 2\hat{j} - 2\hat{k}$, respectively, in a laboratory system of reference. Calculate the position vector of the centre of mass of the system.

Solution: The position vector $\vec{R}$ of the centre of masses is

$$\vec{R} = \frac{m_1 \vec{r}_1 + m_2 \vec{r}_2 + m_3 \vec{r}_3}{m_1 + m_2 + m_3}$$

$$= \frac{1(2\hat{i} + \hat{j} - 4\hat{k}) + 2(\hat{i} + 2\hat{j} - \hat{k}) + 3(3\hat{i} + 2\hat{j} - 2\hat{k})}{1 + 2 + 3}$$

$$= \frac{(13\hat{i} + 11\hat{j} - 12\hat{k})}{6}$$

3.2.1 Velocity of centre of mass

We know that time derivative of displacement is velocity. When position vector depends on time, its differentiation with respect to time gives the velocity. On differentiation of equation (3.24) with respect to time t, the velocity $\vec{V}$ of the centre of mass of the system is

$$\vec{V} = \frac{d\vec{R}}{dt} = \frac{m_1 \vec{v}_1 + m_2 \vec{v}_2 + \ldots + m_n \vec{v}_n}{m_1 + m_2 + \ldots + m_n} \tag{3.25}$$

where

$$\vec{v}_1 = \frac{d\vec{r}_1}{dt}, \qquad \vec{v}_2 = \frac{d\vec{r}_2}{dt}, \qquad \ldots, \qquad \vec{v}_n = \frac{d\vec{r}_n}{dt}$$

are velocities of bodies 1, 2, ..., n, respectively. We know that momentum of a body is the product of its mass and its velocity. In terms of momenta of bodies 1, 2, ..., n, equation (3.25) can be expressed as

$$\vec{V} = \frac{\vec{p}_1 + \vec{p}_2 + \ldots + \vec{p}_n}{m_1 + m_2 + \ldots + m_n} \tag{3.26}$$

where

$$\vec{p}_1 = m_1\vec{v}_1 \qquad \vec{p}_2 = m_2\vec{v}_2 \qquad \ldots \qquad \vec{p}_n = m_n\vec{v}_n$$

are momenta of bodies 1, 2, ..., n, respectively.

Exercise 3.4: Suppose three masses 2, 3 and 4 kg are moving with velocities $\vec{v}_1 = \hat{\imath} + 2\hat{\jmath} + 3\hat{k}$, $\vec{v}_2 = -\hat{\imath} - 3\hat{\jmath} + \hat{k}$ and $\vec{v}_3 = 3\hat{\imath} - 2\hat{\jmath} + 2\hat{k}$, respectively, in a laboratory system of reference. Calculate the velocity of the centre of mass of the system.

Solution: The position vector $\vec{R}$ of the centre of masses is

$$\vec{V} = \frac{m_1\vec{v}_1 + m_2\vec{v}_2 + m_3\vec{v}_3}{m_1 + m_2 + m_3}$$

$$= \frac{2(\hat{\imath} + 2\hat{\jmath} + 3\hat{k}) + 3(-\hat{\imath} - 3\hat{\jmath} + \hat{k}) + 4(3\hat{\imath} - 2\hat{\jmath} + 2\hat{k})}{2 + 3 + 4}$$

$$= \frac{(11\hat{\imath} - 13\hat{\jmath} + 17\hat{k})}{9}$$

Exercise 3.5: Three bodies of masses 2, 3 and 4 kg have the positions $\vec{r}_1 = 5t^2\hat{\imath} + (2t + 1)\hat{\jmath} + (t^3 - t)\hat{k}$, $\vec{r}_2 = (2t^3 - 3)\hat{\imath} - 3t^2\hat{\jmath} + (4t^2 + t)\hat{k}$, $\vec{r}_3 = (t^2 + 1)\hat{\imath} - 2t\hat{\jmath} + (t^2 - 3)\hat{k}$, where t denotes the time. Calculate at $t = 1$ (i) the coordinates of the centre of mass, (ii) velocity of the centre of mass, (iii) total momentum of the system.

Solution: The position vector of the centre of mass is

$$\vec{R} = \frac{m_1\vec{r}_1 + m_2\vec{r}_2 + m_3\vec{r}_3}{m_1 + m_2 + m_3}$$

$$= \frac{\begin{array}{c} 2[5t^2\hat{\imath} + (2t+1)\hat{\jmath} + (t^3 - t)\hat{k}] + 3[(2t^3 - 3)\hat{\imath} - 3t^2\hat{\jmath} \\ +(4t^2 + t)\hat{k}] + 4[(t^2 + 1)\hat{\imath} - 2t\hat{\jmath} + (t^2 - 3)\hat{k}] \end{array}}{2 + 3 + 4}$$

$$= \frac{(6t^3 + 14t^2 - 5)\hat{\imath} - (9t^2 + 4t - 2)\hat{\jmath} + (2t^3 + 16t^2 + t - 12)\hat{k}}{9}$$

The position vector of the centre of mass at $t = 1$ is

$$(\vec{R})_{t=1} = \frac{15\hat{\imath} - 11\hat{\jmath} + 7\hat{k}}{9}$$

Thus, the coordinates of the centre of mass at $t = 1$ are $(15/9, -11/9, 7/9)$.

(ii) The velocity of the centre of mass is

$$\vec{V} = \frac{d\vec{R}}{dt} = \frac{(18t^2 + 28t)\hat{\imath} - (18t + 4)\hat{\jmath} + (6t^2 + 32t + 1)\hat{k}}{9}$$

The velocity of the centre of mass at $t = 1$ is

$$(\vec{V})_{t=1} = \frac{46\hat{\imath} - 22\hat{\jmath} + 39\hat{k}}{9}$$

The magnitude of velocity of the centre of mass at $t = 1$ is

$$(V)_{t=1} = \sqrt{\frac{46^2 + 22^2 + 39^2}{81}} = 7.13$$

(iii) The velocities of the bodies are

$$\vec{v}_1 = 10t\hat{\imath} + 2\hat{\jmath} + (3t^2 - 1)\hat{k}$$

$$\vec{v}_2 = 6t^2\hat{\imath} - 6t\hat{\jmath} + (8t + 1)\hat{k}$$

$$\vec{v}_3 = 2t\hat{\imath} - 2\hat{\jmath} + 2t\hat{k}$$

respectively. The total momentum of the system is

$$\vec{P} = \vec{p}_1 + \vec{p}_2 + \vec{p}_3 = m_1\vec{v}_1 + m_2\vec{v}_2 + m_3\vec{v}_3$$

$$= 2[10t\hat{\imath} + 2\hat{\jmath} + (3t^2 - 1)\hat{k}] + 3[6t^2\hat{\imath} - 6t\hat{\jmath} + (8t + 1)\hat{k})$$

$$+4(2t\hat{\imath} - 2\hat{\jmath} + 2t\hat{k})$$

$$= (18t^2 + 28t)\hat{\imath} - (18t + 4)\hat{\jmath} + (6t^2 + 32t + 1)\hat{k}$$

The total momentum of the system at $t = 1$ is

$$(\vec{P})_{t=1} = 46\hat{\imath} - 22\hat{\jmath} + 39\hat{k}$$

The magnitude of total momentum of the system at $t = 1$ is

$$(P)_{t=1} = \sqrt{46^2 + 22^2 + 39^2} = 60.19$$

3.2.2 Acceleration of centre of mass

On differentiation of equation (3.26) with respect to time, we have the acceleration $\vec{A}$ of the centre of mass of the system as

$$\vec{A} = \frac{d\vec{V}}{dt} = \frac{1}{m_1 + m_2 + \ldots + m_n} \frac{d}{dt}(\vec{p}_1 + \vec{p}_2 + \ldots + \vec{p}_n) \tag{3.27}$$

Thus, the external force $\vec{F}$ acting on the system is

$$\vec{F} = (m_1 + m_2 + \ldots + m_n)\,\vec{A} = \frac{d(\vec{p}_1 + \vec{p}_2 + \ldots + \vec{p}_n)}{dt}$$

When the external force acting on the system is zero, we have

$$\frac{d(\vec{p}_1 + \vec{p}_2 + \ldots + \vec{p}_n)}{dt} = 0$$

or

$$\vec{p}_1 + \vec{p}_2 + \ldots + \vec{p}_n = \text{constant}$$

It is the law of conservation of momentum.

Exercise 3.6: A particle of mass m_1 is moving along x-axis with velocity v_1 and another particle of mass m_2 is stationary. Both particles combine and move with velocity v. Calculate the loss of kinetic energy.

Solution: This is the case of one-dimensional motion along x-axis. On applying the law of conservation of momentum, we have

$$m_1 v_1 = (m_1 + m_2)v \qquad \text{or} \qquad v = \frac{m_1 v_1}{(m_1 + m_2)}$$

This velocity v will remain constant, as no external force is acting on the system of two particles. The kinetic energy of the system before combination is

$$K_i = \frac{1}{2} m_1 v_1^2$$

The kinetic energy of the system after combination is

$$K_f = \frac{1}{2}(m_1 + m_2)v^2$$

Using here the expression for v, we get

$$K_f = \frac{m_1^2 v_1^2}{2(m_1 + m_2)}$$

Change in kinetic energy of the system is

$$\Delta K = K_i - K_f = \frac{1}{2} m_1 v_1^2 - \frac{m_1^2 v_1^2}{2(m_1 + m_2)} = \frac{m_1 m_2 v_1^2}{2(m_1 + m_2)}$$

Since ΔK is positive, we have $K_f < K_i$. Thus, there is loss of kinetic energy. This loss of kinetic energy appears in the internal excitations and heat of the composite system after the collision. When a meteorite m_1 strikes and sticks to the earth m_2, essentially all the kinetic energy of the meteorite appears as heat on the earth.

Exercise 3.7: Two particles of masses m_1 and m_2 are moving along x-axis with velocities v_1 and v_2, respectively, along the same direction. Both particles collide and combine and then move with velocity v. Calculate the loss of kinetic energy.

Solution: This is the case of one-dimensional motion along x-axis. On applying the law of conservation of momentum, we have

$$m_1v_1 + m_2v_2 = (m_1 + m_2)v \qquad \text{or} \qquad v = \frac{m_1v_1 + m_2v_2}{(m_1 + m_2)}$$

This velocity v will remain constant, as no external force is acting on the system of two particles. The kinetic energy of the system before combination is

$$K_i = \frac{1}{2}\,m_1v_1^2 + \frac{1}{2}\,m_2v_2^2$$

The kinetic energy of the system after combination is

$$K_f = \frac{1}{2}(m_1 + m_2)v^2$$

Using expression for v, we get

$$K_f = \frac{m_1^2v_1^2 + m_2^2v_2^2 + 2m_1m_2v_1v_2}{2(m_1 + m_2)}$$

Change in kinetic energy of the system is

$$\Delta K = K_i - K_f = \frac{1}{2}\,m_1v_1^2 + \frac{1}{2}\,m_2v_2^2 - \frac{m_1^2v_1^2 + m_2^2v_2^2 + 2m_1m_2v_1v_2}{2(m_1 + m_2)}$$

$$= \frac{m_1m_2(v_1 - v_2)^2}{2(m_1 + m_2)}$$

Since ΔK is positive, we have $K_f < K_i$. Thus, there is loss of kinetic energy. This loss of kinetic energy appears in the internal excitations and heat of the composite system after the collision.

Exercise 3.8: Two particles of masses m_1 and m_2 are moving along x-axis with velocities v_1 and v_2, respectively, in opposite directions. Both particles collide and combine and then move with velocity v. Calculate the loss of kinetic energy.

Solution: This is the case of one-dimensional motion along x-axis. On applying the law of conservation of momentum, we have

$$m_1v_1 - m_2v_2 = (m_1 + m_2)v \qquad \text{or} \qquad v = \frac{m_1v_1 - m_2v_2}{(m_1 + m_2)}$$

This velocity v will remain constant, as no external force is acting on the system of two particles. The kinetic energy of the system before collision is

$$K_i = \frac{1}{2} m_1 v_1^2 + \frac{1}{2} m_2 v_2^2$$

The kinetic energy of the system after collision is

$$K_f = \frac{1}{2}(m_1 + m_2)v^2$$

Using expression for v, we get

$$K_f = \frac{m_1^2 v_1^2 + m_2^2 v_2^2 - 2m_1 m_2 v_1 v_2}{2(m_1 + m_2)}$$

Change in kinetic energy of the system is

$$\Delta K = K_i - K_f = \frac{1}{2} m_1 v_1^2 + \frac{1}{2} m_2 v_2^2 - \frac{m_1^2 v_1^2 + m_2^2 v_2^2 - 2m_1 m_2 v_1 v_2}{2(m_1 + m_2)}$$

$$= \frac{m_1 m_2 (v_1 + v_2)^2}{2(m_1 + m_2)}$$

Since ΔK is positive, we have $K_f < K_i$. Thus, there is loss of kinetic energy. This loss of kinetic energy appears in the internal excitations and heat of the composite system after the collision.

Note: All the treatments we are discussing are in the laboratory frame of reference.

Exercise 3.9: Two bodies 1 and 2 of masses m_1 and m_2 are moving with velocities $\vec{v}_1$ and $\vec{v}_2$, respectively, in a laboratory frame of reference. Show that in a centre-of-mass system, the momenta of the bodies are equal and opposite to each other.

Solution: When two bodies 1 and 2 of masses m_1 and m_2 are moving with velocities $\vec{v}_1$ and $\vec{v}_2$, respectively, in a laboratory frame of reference, their momenta are

$$\vec{p}_1 = m_1 \vec{v}_1 \quad \text{and} \quad \vec{p}_2 = m_2 \vec{v}_2$$

These two momenta are independent of each other. The velocity of the centre-of-mass system of two bodies is

$$\vec{V} = \frac{m_1 \vec{v}_1 + m_2 \vec{v}_2}{m_1 + m_2}$$

The velocity of particle 1 in the centre-of-mass system is

$$\vec{u}_1 = \vec{v}_1 - \vec{V} = \vec{v}_1 - \frac{m_1 \vec{v}_1 + m_2 \vec{v}_2}{m_1 + m_2} = \frac{m_2(\vec{v}_1 - \vec{v}_2)}{m_1 + m_2}$$

Similarly, the velocity of particle 2 in the centre-of-mass system is

$$\vec{u}_2 = \vec{v}_2 - \vec{V} = \vec{v}_2 - \frac{m_1\vec{v}_1 + m_2\vec{v}_2}{m_1 + m_2} = \frac{m_1(\vec{v}_2 - \vec{v}_1)}{m_1 + m_2}$$

The momenta of the particles 1 and 2, respectively, in the centre-of-mass system are

$$\vec{p}\,'_1 = m_1\vec{u}_1 = \frac{m_1 m_2(\vec{v}_1 - \vec{v}_2)}{m_1 + m_2}$$

and

$$\vec{p}\,'_2 = m_2\vec{u}_2 = \frac{m_1 m_2(\vec{v}_2 - \vec{v}_1)}{m_1 + m_2}$$

showing that $\vec{p}\,'_1 = -\vec{p}\,'_2$. It shows that the momenta of the particles 1 and 2 are equal and opposite to each other in the centre-of-mass system.

Note: While the momenta of two particles in the laboratory system of reference had no relation between them, in the centre-of-mass system the momenta of the particles are equal and opposite to each other.

Exercise 3.10: Two bodies 1 and 2 of masses 0.3 kg and 0.4 kg are moving with velocities $\vec{v}_1 = 2\hat{\imath} - 3\hat{\jmath} + 4\hat{k}$ and $\vec{v}_2 = 2\hat{\imath} - 2\hat{\jmath} + 3\hat{k}$, respectively, in a laboratory system of reference. Show that in the laboratory system of reference, the momenta of the bodies have no relation whereas in the centre-of-mass system, their momenta are equal and opposite to each other.

Solution: The bodies 1 and 2 of masses $m_1 = 0.3$ kg and $m_2 = 0.4$ kg are moving with velocities $\vec{v}_1 = 2\hat{\imath} - 3\hat{\jmath} + 4\hat{k}$ and $\vec{v}_2 = 2\hat{\imath} - 2\hat{\jmath} + 3\hat{k}$, respectively, in a laboratory system of reference. In the laboratory system of reference, momentum of body 1 is

$$\vec{p}_1 = m_1\vec{v}_1 = 0.3(2\hat{\imath} - 3\hat{\jmath} + 4\hat{k}) = 0.6\hat{\imath} - 0.9\hat{\jmath} + 1.2\hat{k}$$

and of body 2 is

$$\vec{p}_2 = m_2\vec{v}_2 = 0.4(2\hat{\imath} - 2\hat{\jmath} + 3\hat{k}) = 0.8\hat{\imath} - 0.8\hat{\jmath} + 1.2\hat{k}$$

showing that $\vec{p}_1$ and $\vec{p}_2$ are independent of each other. The velocity of the centre of mass is

$$\vec{V} = \frac{m_1\vec{v}_1 + m_2\vec{v}_2}{m_1 + m_2}$$

$$= \frac{0.3(2\hat{\imath} - 3\hat{\jmath} + 4\hat{k}) + 0.4(2\hat{\imath} - 2\hat{\jmath} + 3\hat{k})}{0.3 + 0.4}$$

$$= \frac{14}{7}\hat{\imath} - \frac{17}{7}\hat{\jmath} + \frac{24}{7}\hat{k}$$

In the centre-of-mass system, the velocity of body 1 is

$$\vec{u}_1 = \vec{v}_1 - \vec{V} = 2\hat{i} - 3\hat{j} + 4\hat{k} - \left[\frac{14}{7}\,\hat{i} - \frac{17}{7}\,\hat{j} + \frac{24}{7}\,\hat{k}\right]$$

$$= -\frac{4}{7}\hat{j} + \frac{4}{7}\hat{k}$$

and the velocity of body 2 is

$$\vec{u}_2 = \vec{v}_2 - \vec{V} = 2\hat{i} - 2\hat{j} + 3\hat{k} - \left[\frac{14}{7}\,\hat{i} - \frac{17}{7}\,\hat{j} + \frac{24}{7}\,\hat{k}\right]$$

$$= \frac{3}{7}\,\hat{j} - \frac{3}{7}\,\hat{k}$$

In the centre-of-mass system, momentum of body 1 is

$$\vec{p}\,'_1 = m_1\vec{u}_1 = 0.3\left[-\frac{4}{7}\,\hat{j} + \frac{4}{7}\,\hat{k}\right] = -\frac{1.2}{7}\,\hat{j} + \frac{1.2}{7}\,\hat{k}$$

and of body 2 is

$$\vec{p}\,'_2 = m_2\vec{u}_2 = 0.4\left[\frac{3}{7}\,\hat{j} - \frac{3}{7}\,\hat{k}\right] = \frac{1.2}{7}\,\hat{j} - \frac{1.2}{7}\,\hat{k}$$

showing that $\vec{p}\,'_1 = -\vec{p}\,'_2$

3.3 Angular momentum and torque

Suppose a body of mass m is moving with velocity $\vec{v}$, its momentum is $\vec{p} = m\vec{v}$. The body is at a point A whose position vector is $\vec{r}$, as shown in the following figure, where both $\vec{p}$ and $\vec{r}$ lie in the plane of the paper. Let us consider an axis perpendicular to the plane of the paper and passing through a point O, referred to as the point of reference. Such axis is generally referred to as the axis of rotation. The angular momentum $\vec{J}$ of the body about O is expressed as

$$\vec{J} = \vec{r} \times \vec{p}$$

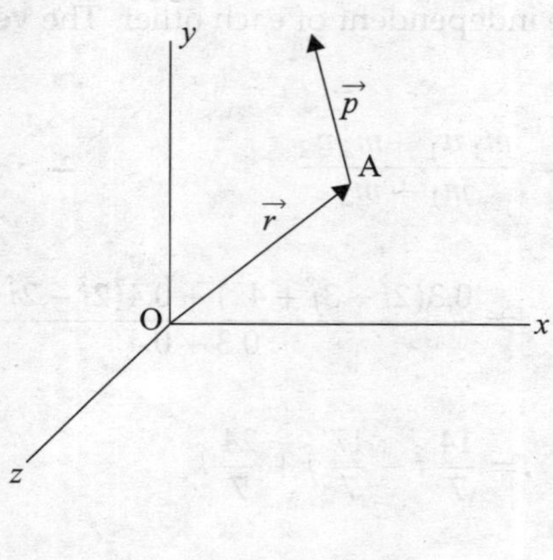

Suppose an external force $\vec{F}$ is applied on a body placed at a point A whose position vector is $\vec{r}$, as shown in the following figure, where both $\vec{p}$ and $\vec{r}$ lie in the plane of the paper. Let us consider an axis perpendicular to the plane of the paper and passing through a point O, referred to as the point of reference. Such axis is generally referred to as the axis of rotation. The torque $\vec{\tau}$ of the force $\vec{F}$ about O is expressed as

$$\vec{\tau} = \vec{r} \times \vec{F} \tag{3.28}$$

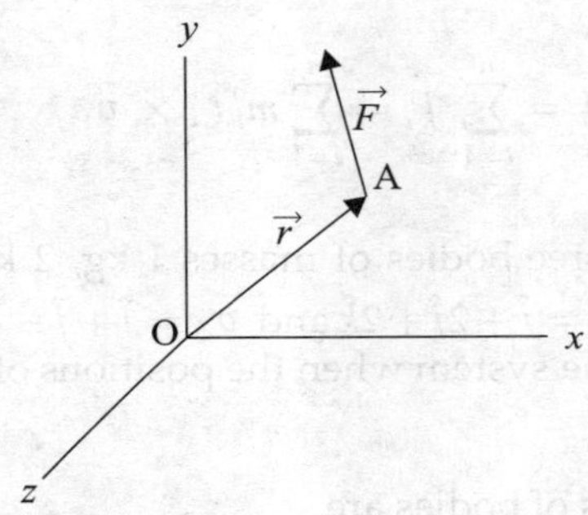

On differentiating equation (3.28) with respect to time, we get

$$\frac{d\vec{J}}{dt} = \frac{d\vec{r}}{dt} \times \vec{p} + \vec{r} \times \frac{d\vec{p}}{dt}$$

Using

$$\vec{v} = \frac{d\vec{r}}{dt} \quad \text{and} \quad \vec{F} = \frac{d\vec{p}}{dt}$$

we get

$$\frac{d\vec{J}}{dt} = \vec{v} \times m\vec{v} + \vec{r} \times \vec{F} = \vec{\tau}$$

Here, we have used the relation $\vec{v} \times \vec{v} = 0$. Thus, we have

$$\vec{\tau} = \frac{d\vec{J}}{dt}$$

The rate of change of angular momentum of a body is equal to the external torque applied on the body.[1] It shows that when $\vec{\tau} = 0$, the angular momentum $\vec{J}$ is conserved. Thus, the statement for conservation of angular momentum is that when no external torque is applied, the angular momentum is conserved. Note that for the conservation of angular momentum, the orbit of body need not to be close.

[1]This relation is very similar to the equation in the context of Newton's laws of motion.

3.3.1 Angular momentum of a system of masses

Let us consider a system consisting n bodies. Suppose $\vec{r}_i$ denotes position vector of ith body of mass m_i moving with velocity $\vec{v}_i$. The angular momentum of the ith body about the origin of coordinate system is

$$\vec{J}_i = m_i \vec{r}_i \times \vec{v}_i$$

Thus, total angular momentum $\vec{J}$ of the system is

$$\vec{J} = \sum_{i=1}^{n} \vec{J}_i = \sum_{i=1}^{n} m_i \vec{r}_i \times \vec{v}_i$$

Exercise 3.11: In a system, three bodies of masses 1 kg, 2 kg and 1 kg are moving with velocities $\vec{v}_1 = 2\hat{i} - 3\hat{j} + \hat{k}$, $\vec{v}_2 = \hat{i} + 2\hat{j} + 2\hat{k}$ and $\vec{v}_3 = \hat{i} + \hat{j} - \hat{k}$ m/s, respectively. Calculate total angular momentum of the system when the positions of the masses are (1, 1, 2), (1, 2, 3) and (1, 4, 2), respectively.

Solution: The position vectors of bodies are

$$\vec{r}_1 = \hat{i} + \hat{j} + 2\hat{k} \qquad \vec{r}_2 = \hat{i} + 2\hat{j} + 3\hat{k} \qquad \vec{r}_3 = \hat{i} + 4\hat{j} + 2\hat{k}$$

We have

$$\vec{r}_1 \times \vec{v}_1 = \begin{vmatrix} \hat{i} & \hat{j} & \hat{k} \\ 1 & 1 & 2 \\ 2 & -3 & 1 \end{vmatrix} = 7\hat{i} + 3\hat{j} - 5\hat{k}$$

$$\vec{r}_2 \times \vec{v}_2 = \begin{vmatrix} \hat{i} & \hat{j} & \hat{k} \\ 1 & 2 & 3 \\ 1 & 2 & 2 \end{vmatrix} = -2\hat{i} + \hat{j}$$

$$\vec{r}_3 \times \vec{v}_3 = \begin{vmatrix} \hat{i} & \hat{j} & \hat{k} \\ 1 & 4 & 2 \\ 1 & 1 & -1 \end{vmatrix} = -6\hat{i} + 3\hat{j} - 3\hat{k}$$

The total angular momentum of the system is

$$\vec{J} = \sum_{i=1}^{3} m_i \vec{r}_i \times \vec{v}_i$$

$$= 1(7\hat{i} + 3\hat{j} - 5\hat{k}) + 2(-2\hat{i} + \hat{j}) + 1(-6\hat{i} + 3\hat{j} - 3\hat{k}) = 3\hat{i} + 8\hat{j} - 8\hat{k}$$

The magnitude of total angular momentum is

$$J = |\vec{J}| = \sqrt{9 + 64 + 64} = \sqrt{137}$$

3.3.2 Angular momentum about the centre of mass

Let us consider a system consisting n bodies. Suppose $\vec{r}_i$ denotes position vector of ith body of mass m_i moving with velocity $\vec{v}_i$. The angular momentum of the ith body about the origin of coordinate system is

$$\vec{J}_i = m_i \vec{r}_i \times \vec{v}_i$$

Thus, total angular momentum $\vec{J}$ of the system is

$$\vec{J} = \sum_{i=1}^{n} \vec{J}_i = \sum_{i=1}^{n} m_i \vec{r}_i \times \vec{v}_i \tag{3.29}$$

The position vector $\vec{R}$ of centre of mass is expressed as

$$\vec{R} = \sum_{i=1}^{n} m_i \vec{r}_i \, / \sum_{i=1}^{n} m_i$$

On adding and subtracting the term

$$\sum_{i=1}^{n} m_i \vec{R} \times \vec{v}_i$$

to right side of equation (3.29), we get

$$\begin{aligned}
\vec{J} &= \sum_{i=1}^{n} m_i \vec{r}_i \times \vec{v}_i - \sum_{i=1}^{n} m_i \vec{R} \times \vec{v}_i + \sum_{i=1}^{n} m_i \vec{R} \times \vec{v}_i \\
&= \sum_{i=1}^{n} m_i (\vec{r}_i - \vec{R}) \times \vec{v}_i + \vec{R} \times \sum_{i=1}^{n} m_i \vec{v}_i \\
&= \vec{J}_{\text{cm}} + \vec{R} \times \vec{P}
\end{aligned}$$

where

$$\vec{J}_{\text{cm}} = \sum_{i=1}^{n} m_i (\vec{r}_i - \vec{R}) \times \vec{v}_i \qquad \text{and} \qquad \vec{P} = \sum_{i=1}^{n} m_i \vec{v}_i$$

Here, $\vec{J}_{\text{cm}}$ is known as the angular momentum about the centre of mass and $\vec{P}$ is the total momentum of the system. Notice that in $\vec{J}_{\text{cm}}$, the positions of bodies are measured relative to the centre of mass. The term $\vec{R} \times \vec{P}$ is the angular momentum due to motion of centre of mass about the origin of the coordinate system. The external torque $\vec{\tau}$ acting on the system is

$$\vec{\tau} = \frac{d\vec{J}}{dt}$$

When we choose origin of the coordinate system at the centre of mass, then $\vec{R} = 0$ and we have

$$\vec{J} = \vec{J}_{\text{cm}} \quad \text{and} \quad \vec{\tau} = \frac{d\vec{J}_{\text{cm}}}{dt}$$

Now, when no external torque acts on the system, the angular momentum about the centre of mass is conserved.

Exercise 3.12: In a system, two bodies of masses 2 kg and 1 kg are moving with velocities $\vec{v}_1 = \hat{i} + 2\hat{j} + \hat{k}$ and $\vec{v}_2 = 2\hat{i} - \hat{j} + 2\hat{k}$ m/s, respectively. Calculate total angular momentum of the system about its centre of mass when the positions of the masses are (1, 3, 2) and (2, 1, 3), respectively.

Solution: The position vectors of bodies are

$$\vec{r}_1 = \hat{i} + 3\hat{j} + 2\hat{k} \quad \text{and} \quad \vec{r}_2 = 2\hat{i} + \hat{j} + 3\hat{k}$$

The position vector $\vec{R}$ of centre of mass is

$$\vec{R} = \frac{m_1 \vec{r}_1 + m_2 \vec{r}_2}{m_1 + m_2}$$

$$= \frac{2(\hat{i} + 3\hat{j} + 2\hat{k}) + 1(2\hat{i} + \hat{j} + 3\hat{k})}{2+1} = \frac{4\hat{i} + 7\hat{j} + 7\hat{k}}{3}$$

So, the position vectors of bodies relative to the centre of mass are

$$\vec{r}\,'_1 = \vec{r}_1 - \vec{R} = \hat{i} + 3\hat{j} + 2\hat{k} - \frac{4\hat{i} + 7\hat{j} + 7\hat{k}}{3} = \frac{-\hat{i} + 2\hat{j} - \hat{k}}{3}$$

and

$$\vec{r}\,'_2 = \vec{r}_2 - \vec{R} = 2\hat{i} + \hat{j} + 3\hat{k} - \frac{4\hat{i} + 7\hat{j} + 7\hat{k}}{3} = \frac{2\hat{i} - 4\hat{j} + 2\hat{k}}{3}$$

We have

$$\vec{r}\,'_1 \times \vec{v}_1 = \begin{vmatrix} \hat{i} & \hat{j} & \hat{k} \\ -1/3 & 2/3 & -1/3 \\ 1 & 2 & 1 \end{vmatrix} = \tfrac{4}{3}\,\hat{i} - \tfrac{4}{3}\,\hat{k}$$

and

$$\vec{r}\,'_2 \times \vec{v}_2 = \begin{vmatrix} \hat{i} & \hat{j} & \hat{k} \\ 2/3 & -4/3 & 2/3 \\ 2 & -1 & 2 \end{vmatrix} = -2\,\hat{i} + 2\,\hat{k}$$

The total angular momentum of the system about the centre of mass is

$$\vec{J}_{\text{cm}} = \sum_{i=1}^{2} m_i\, \vec{r}\,'_i \times \vec{v}_i$$

$$= 2(\frac{4}{3}\,\hat{\imath} - \frac{4}{3}\,\hat{k}) + 1(-2\hat{\imath} + 2\hat{k}) = \frac{2}{3}\,\hat{\imath} - \frac{2}{3}\,\hat{k}$$

The magnitude of total momentum is

$$J_{\text{cm}} = |\,\vec{J}_{\text{cm}}\,| = \sqrt{\frac{4}{9} + \frac{4}{9}} = \sqrt{\frac{8}{9}}$$

3.3.3 Total torque due to internal forces

Let us consider a system consisting n particles. The particles interact with one another. Suppose $\vec{r}_i$ denotes position vector of ith particle. If $\vec{F}_i$ is total force acting on the ith particle, the torque acting on the ith particle about the origin of coordinate system is

$$\vec{\tau}_i = \vec{r}_i \times \vec{F}_i$$

The force $\vec{F}_i$ consists of two parts (i) external force $\vec{f}_i$ and (ii) internal force[2] due to all other particles. That is

$$\vec{F}_i = \vec{f}_i + \sum_{j=1}^{n}{}' \vec{f}_{ij}$$

Here, $\vec{f}_{ij}$ denotes force on the ith particle due to jth particle and prime over summation sign denotes that sum is taken for all particles except when $j = i$, as a particle cannot apply force on itself. Thus, total torque acting on the system is

$$\vec{\tau} = \sum_{i=1}^{n} \vec{\tau}_i = \sum_{i=1}^{n} \vec{r}_i \times \Big(\vec{f}_i + \sum_{j=1}^{n}{}' \vec{f}_{ij} \Big)$$

$$= \sum_{i=1}^{n} \vec{r}_i \times \vec{f}_i + \sum_{i=1}^{n} \sum_{j=1}^{n}{}' \vec{r}_i \times \vec{f}_{ij}$$

Here, the second term is the torque due to internal forces

$$\vec{\tau}_{\text{int}} = \sum_{i=1}^{n} \sum_{j=1}^{n}{}' \vec{r}_i \times \vec{f}_{ij}$$

Using the fact $\vec{f}_{ij} = -\vec{f}_{ji}$, and doing simple rearrangement, we can write[3]

$$\vec{\tau}_{\text{int}} = \sum_{i=1}^{n-1} \sum_{j+1}^{n} (\vec{r}_i - \vec{r}_j) \times \vec{f}_{ij}$$

[2]The forces between constituent particles of a system are known as internal forces.

[3]For understanding this relation, let us take a case of four particles and we have

$$\vec{\tau}_{\text{int}} = \sum_{i=1}^{4} \sum_{j=1}^{4}{}' \vec{r}_i \times \vec{f}_{ij} = \vec{r}_1 \times \vec{f}_{12} + \vec{r}_1 \times \vec{f}_{13} + \vec{r}_1 \times \vec{f}_{14}$$
$$+ \vec{r}_2 \times \vec{f}_{21} + \vec{r}_2 \times \vec{f}_{23} + \vec{r}_2 \times \vec{f}_{24} ++ \vec{r}_3 \times \vec{f}_{31} + \vec{r}_3 \times \vec{f}_{32}$$

Since the separation between the ith and jth particles $(\vec{r}_i - \vec{r}_j)$ is parallel to the force $\vec{f}_{ij}$, we have

$$\vec{\tau}_{\text{int}} = 0$$

showing that the torque due to internal forces in a system is zero.

3.4 Rotation along a circular path

For motion along a circular path (Figure 3.5), the direction of position vector $\vec{r}$ and linear momentum $\vec{p}$ are always perpendicular to each other

$$\vec{J} = \vec{r} \times \vec{p} = rp\,\hat{n}$$

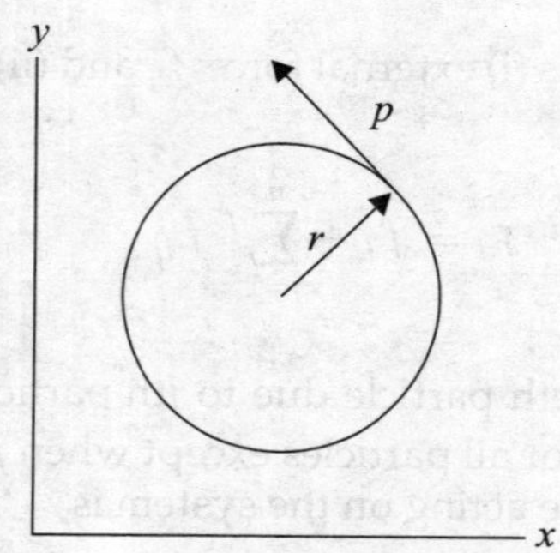

Figure 3.5: Motion along a circular path.

where $\hat{n}$ is a unit vector perpendicular to the plane containing $\vec{r}$ and $\vec{p}$ both. The magnitude of angular momentum is

$$J = rp = rmv$$

Using relation between angular velocity ω and velocity v, $v = r\omega$, we get

$$J = mr^2\omega$$

The moment of inertia of the body about the axis of rotation is $I = mr^2$ and therefore, we have

---

$$+\,\vec{r}_3 \times \vec{f}_{34} + \vec{r}_4 \times \vec{f}_{41} + \vec{r}_4 \times \vec{f}_{42} + \vec{r}_4 \times \vec{f}_{43}$$

Using the fact $\vec{f}_{ij} = -\vec{f}_{ji}$, we have

$$\vec{\tau}_{\text{int}} = (\vec{r}_1 - \vec{r}_2)\times \vec{f}_{12} + (\vec{r}_1 - \vec{r}_3)\times \vec{f}_{13} + (\vec{r}_1 - \vec{r}_4)\times \vec{f}_{14} + (\vec{r}_2 - \vec{r}_3)\times \vec{f}_{23}$$

$$+(\vec{r}_2 - \vec{r}_4)\times \vec{f}_{24} + (\vec{r}_3 - \vec{r}_4)\times \vec{f}_{34} = \sum_{i=1}^{3}\sum_{j=i+1}^{4} (\vec{r}_i - \vec{r}_j)\times \vec{f}_{ij}$$

$$J = I\omega$$

On comparing this expression with the relation $p = mv$, the moment of inertia I may be called as the rotational mass. Suppose a body revolving about the axis of rotation is brought closer to the axis, the moment of inertia of the system decreases, as the distance from the axis of rotation is decreased. For conservation of angular momentum, the angular velocity ω of the system will increase.

3.5 Formation of a galaxy

For understanding formation of a galaxy, let us consider a body of mass m revolving about an axis which is at a distance r_1 from the body (Figure 3.6). The moment of inertia of the body about the axis of rotation is

$$I_1 = mr_1^2 \tag{3.30}$$

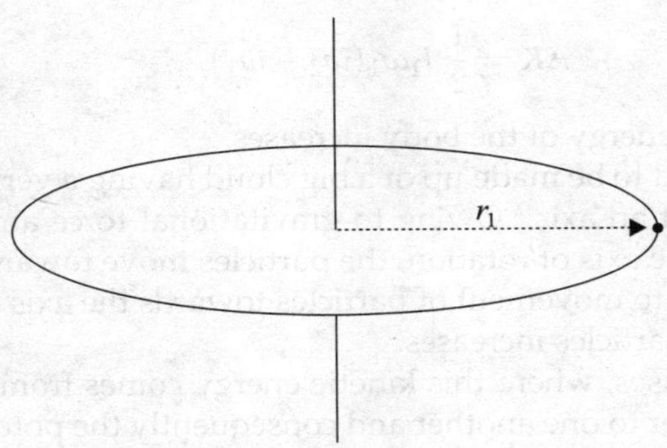

Figure 3.6: Rotation of a body of mass m about the axis of rotation.

Suppose the angular velocity of revolution about the axis of rotation is ω_1. Then the magnitude of angular momentum J_1 of the body about the axis of rotation is

$$J_1 = I_1\omega_1$$

and angular kinetic energy is

$$K_1 = \frac{1}{2} I_1\omega_1^2$$

Suppose the body moved from the distance r_1 to a distance r_2 from the axis of rotation, such that $r_2 < r_1$. At this position, the moment of inertia of the body about the axis of rotation is

$$I_2 = mr_2^2 \tag{3.31}$$

Since $r_2 < r_1$, from equations (3.30) and (3.31), we have $I_2 < I_1$. Suppose the angular velocity of revolution about the axis of rotation is ω_2. Then the magnitude of angular momentum J_2 of the system about the axis of rotation is

$$J_2 = I_2\omega_2$$

and angular kinetic energy is

$$K_2 = \frac{1}{2}\, I_2\omega_2^2$$

When no external torque is applied for moving the body from one position to another, the angular momentum is conserved and we have

$$J_1 = J_2 \qquad \text{and} \qquad I_1\omega_1 = I_2\omega_2 \tag{3.32}$$

Since $I_2 < I_1$, we have $\omega_2 > \omega_1$. It shows that when in absence of any external torque, the body comes closer to the axis of rotation, the angular velocity of the body increases. The change in kinetic energy is

$$\Delta K = K_2 - K_1 = \frac{1}{2}\, I_2\omega_2^2 - \frac{1}{2}\, I_1\omega_1^2 \tag{3.33}$$

Using equation (3.32) in (3.33), we have

$$\Delta K = \frac{1}{2}\, I_1\omega_1(\omega_2 - \omega_1)$$

It shows that the kinetic energy of the body increases.

A galaxy is considered to be made up of a big cloud having a very large number of tiny particles revolving about an axis. Owing to gravitational force among various particles and large density near the axis of rotation, the particles move towards the axis of rotation. As discussed above, due to movement of particles towards the axis of rotation, the kinetic energy of the system of particles increases.

Now, the question arises, where this kinetic energy comes from. In the movement of particles, they come closer to one another and consequently the potential energy

$$\frac{Gm_1m_2}{r}$$

decreases. Hence, the system loses gravitational energy. That is in order to increase the separation between particles, we have to do work. This decrease in gravitational energy appears as the increase of kinetic energy of the system.

3.5.1 Mass of a galaxy

A galaxy rotates about its own axis and most of its mass is in the central part of the galaxy. Suppose we have a star of mass M revolving around the centre of galaxy with radius r and velocity v. The force of attraction between the star and the galaxy is balanced by the centrifugal force. The equilibrium equation is

$$\frac{GMM_g}{r^2} = \frac{Mv^2}{r} \qquad \text{or} \qquad v^2r = GM_g \tag{3.34}$$

Here, M_g stands for the mass of galaxy and G is the universal gravitational constant. Further, we have

$$T = \frac{2\pi r}{v} \qquad \text{or} \qquad v = \frac{2\pi r}{T} \tag{3.35}$$

Using equation (3.35) in (3.34), we have

$$T^2 = \frac{4\pi^2}{GM} r^3 \qquad \text{or} \qquad M = \frac{4\pi^2}{G} \frac{r^3}{T^2}$$

Thus, by knowing the distance r of sun from the centre of galaxy and the time period T of revolution of the sun around the centre of galaxy, we can calculate mass of the galaxy.

3.6 Solar system

We know that universe is made up of a large number of galaxies. Our galaxy is named as the Milky Way. Each galaxy has billions of stars. The name of our star is the sun. The sun has eight planets revolving around it. These planets are Mercury, Venus, Earth, Mars, Jupiter, Saturn, Uranus and Neptune. (Earlier, we had one more planet, Pluto, which has been now excluded from the list of planets.) Each of the planets revolves around the sun and spins about its own axis. Thus, each planet has two angular motions: (i) revolution around the sun, and (ii) rotation about its own axis.

Although there are other bodies (asteroids and those lying beyond Neptune) also revolving around the sun, in the present discussion we are including only eight planets, as they are the main contributors to the angular momentum of solar system. There are also satellites revolving around their respective planets. Each of these satellites rotates (spins) about its own axis as well. These satellites also are not included in the present discussion, as their contribution to the angular momentum of the solar system is negligibly small.

Although the orbit of revolution of a planet around the sun is elliptical in shape, for our discussion, we shall consider it as circular. Some physical parameters for sun and its eight planets are given in Table 3.1.

Table 3.1: Parameters for sun and planets

Object	Mean distance (m)	Mass (kg)	Orbital period (s)	Radius (m)	Rotational period (s)
Sun		1.99×10^{30}		6.96×10^{8}	2.16×10^{6}
Mercury	5.8×10^{10}	3.30×10^{23}	7.6×10^{6}	2.4×10^{6}	5.1×10^{6}
Venus	1.1×10^{11}	4.87×10^{24}	1.9×10^{7}	6.1×10^{6}	2.1×10^{7}
Earth	1.5×10^{11}	5.97×10^{24}	3.2×10^{7}	6.4×10^{6}	8.6×10^{4}
Mars	2.3×10^{11}	6.42×10^{23}	5.9×10^{7}	3.4×10^{6}	8.9×10^{4}
Jupiter	7.8×10^{11}	1.90×10^{27}	3.7×10^{8}	7.1×10^{7}	3.6×10^{4}
Saturn	1.4×10^{12}	5.68×10^{26}	9.3×10^{8}	6.0×10^{7}	3.8×10^{4}
Uranus	2.9×10^{12}	8.68×10^{25}	2.6×10^{9}	2.6×10^{7}	6.2×10^{4}
Neptune	4.5×10^{12}	1.02×10^{26}	5.2×10^{9}	2.5×10^{7}	5.8×10^{4}

Suppose m is mass of a satellite revolving around the sun in a circular orbit of radius r with time period T. The gravitational force of attraction between a planet and the sun is balanced by the centrifugal force, and the relation for equilibrium is

$$\frac{GMm}{r^2} = \frac{mv^2}{r} \qquad \text{or} \qquad v^2 r = GM \tag{3.36}$$

Here, M stands for the mass of the sun and G is the universal gravitational constant. Further, we have

$$T = \frac{2\pi r}{v} \quad \text{or} \quad v = \frac{2\pi r}{T} \tag{3.37}$$

Using equation (3.37) in (3.36), we have

$$T^2 = \frac{4\pi^2}{GM} r^3 \quad \text{or} \quad M = \frac{4\pi^2}{G}\frac{r^3}{T^2}$$

Thus, by knowing distance r of a planet from the sun and the time period T of revolution of the planet around the sun, we can calculate mass of the sun.

Exercise 3.13: Distance of earth from the sun is 1.5×10^8 km and it completes one revolution around the sun in 365.25 days. Calculate mass of the sun. Given $G = 6.67 \times 10^{-11}$ N-m^2/kg^2.

Solution: Distance of earth from the sun is

$$r = 1.5 \times 10^8 \text{ km} = 1.5 \times 10^{11} \text{ m}$$

The time period of revolution of earth around the sun is

$$T = 365.25 \text{ days} = 365.25 \times 24 \times 60 \times 60 = 3.16 \times 10^7 \text{ s}$$

The mass of the sun is

$$M = \frac{4\pi^2}{G}\frac{r^3}{T^2} = \frac{4 \times (3.14)^2 \times (1.5 \times 10^{11})^2}{6.67 \times 10^{-11} \times (3.16 \times 10^7)^3} = 2.0 \times \times 10^{30} \text{ kg}$$

3.6.1 Angular momentum for revolution and rotation

The angular velocity is

$$\omega = \frac{2\pi}{T} \tag{3.38}$$

and relation between velocity v and angular velocity ω is

$$v = r\omega \tag{3.39}$$

For a circular orbit, the direction of radial vector $\vec{r}$ and that of momentum $\vec{p}$ are perpendicular to each other, and therefore the angular momentum $\vec{J}$ is

$$\vec{J} = \vec{r} \times \vec{p} = rp\,\hat{n}$$

where r and p are magnitudes of $\vec{r}$ and $\vec{p}$, respectively, and $\hat{n}$ is the unit vector perpendicular to the plane containing $\vec{r}$ and $\vec{p}$ both. Thus, we have the magnitude of $\vec{J}$ as

$$J = rp = mrv \tag{3.40}$$

Using equation (3.39) in (3.40), we get

$$J = mr^2\omega$$

Using equation (3.38) here, we get

$$J = \frac{2\pi mr^2}{T} \tag{3.41}$$

This is the expression for angular momentum of a body of mass m revolving in a circular orbit of radius r with time period T.

Let us look for the rotation of a body about its own axis. Here, we shall assume that the body is spherical in shape with radius R and rotating (spinning) with time period T'. For an axis passing through the centre of the body, if we consider that mass m is placed at a distance R and the velocity of the mass is v', the angular momentum $\vec{J}'$ for rotating motion is

$$\vec{J}' = \vec{R} \times \vec{p}' = R\, p'\, \hat{n}'$$

where R and p' are magnitudes of $\vec{R}$ and $\vec{p}'$, respectively, and $\hat{n}'$ is the unit vector perpendicular to the plane containing $\vec{R}$ and $\vec{p}'$ both. Thus, we have

$$J' = Rp' = Rmv'$$

Using the relation $v' = R\omega'$, where ω' is angular velocity of rotation, we have

$$J' = mR^2\omega' = \frac{2\pi mR^2}{T'}$$

where T' denotes the time period of rotation of planet. In this presentation, we have assumed that the mass m is situated at the surface of the body. Since the mass is distributed throughout the body, this effect can be accounted for by replacing mass m by $2m/5$, and therefore, we have

$$J' = \frac{4\pi mR^2}{5T'} \tag{3.42}$$

For discussion of angular momentum of solar system, the sun is considered stationary. However, the sun still has rotational motion about its own axis. The sun does not rotate as a solid body; its various segments rotate with different velocities. Its equator rotates with the time period of 25 days and the rotation period increases up to 35 days near the poles. In our discussion, we are taking 25 days for one rotation of the sun. Thus, rotational angular momentum of the sun is

$$J' = \frac{4\pi mR^2}{5T'} = \frac{4\pi \times 1.99 \times 10^{30} \times (6.96 \times 10^8)^2}{5 \times 2.16 \times 10^6}$$

$$= 1.12 \times 10^{42}\ \text{kg-m}^2/\text{s}$$

For the data given in Table 3.1, we have calculated J and J' with the help of equations (3.41) and (3.42) and the calculated values are given in Table 3.2.

Table 3.2: Value of J and J'

Object	J (kg-m^2/s)	J' (kg-m^2/s)	%
Sun		1.12×10^{42}	3.442
Mercury	9.17×10^{38}	9.68×10^{29}	0.003
Venus	1.95×10^{40}	2.17×10^{31}	0.060
Earth	2.64×10^{40}	7.14×10^{33}	0.081
Mars	3.60×10^{39}	2.09×10^{32}	0.011
Jupiter	1.96×10^{43}	6.68×10^{38}	60.246
Saturn	7.52×10^{42}	3.35×10^{38}	23.084
Uranus	1.76×10^{42}	2.38×10^{36}	5.414
Neptune	2.49×10^{42}	2.76×10^{36}	7.659
sum =	3.14×10^{43}	1.12×10^{42}	

3.7 Angular momentum and kinetic energy for rotational motion

Suppose a particle of mass m is moving with instantaneous velocity $\vec{v}$ along a curved path with angular velocity ω. If $\vec{r}$ be the position vector of the particle and both $\vec{v}$ and $\vec{r}$ lie in the plane of the paper, the direction of angular velocity $\vec{\omega}$ is perpendicular to the plane of the paper and is in the outward direction. Then, we have

$$\vec{v} = \vec{\omega} \times \vec{r} \tag{3.43}$$

The momentum of the particle is

$$\vec{p} = m\vec{v} = m\vec{\omega} \times \vec{r}$$

The angular momentum of the particle is

$$\vec{J} = \vec{r} \times \vec{p} = m\vec{r} \times (\vec{\omega} \times \vec{r})$$

$$= m\,\vec{\omega}\,(\vec{r}\,.\,\vec{r}) - \vec{r}\,(\vec{r}\,.\,\vec{\omega})$$

Since the directions of ω and $\vec{r}$ are perpendicular to each other, we have $\vec{r}\,.\,\omega = 0$ and therefore, we get[4]

$$\vec{J} = mr^2\,\vec{\omega} = I\,\vec{\omega}$$

where $I = mr^2$ is the moment of inertia of the particle about the axis of rotation. The kinetic energy of the particle is[5]

$$K = \frac{1}{2}\,m\vec{v}\,.\,\vec{v} = \frac{1}{2}\,m(\vec{\omega} \times \vec{r}).(\vec{\omega} \times \vec{r})$$

[4]This expression is similar to the relation $\vec{p} = m\,\vec{v}$, showing that the moment of inertia I may be called the angular mass.

[5]This expression is similar to the relation $E = mv^2/2$, showing that the moment of inertia I may be called the angular mass.

$$= \frac{1}{2}\, m |\vec{\omega} \times \vec{r}\,|^2 = \frac{1}{2}\, mr^2\omega^2 = \frac{1}{2}\, I\omega^2$$

where $I = mr^2$ is the moment of inertia of the particle about the axis of rotation.

3.7.1 Case of a body

Let us now consider a body having a number of particles. Suppose m_i and $\vec{r}_i$ are, respectively, the mass and position vector of ith particle of the body. If the body is rotating with angular velocity $\vec{\omega}$ about an axis, the velocity of the ith particle is

$$\vec{v}_i = \vec{\omega} \times \vec{r}_i$$

The angular momentum of the ith particle is

$$\vec{J}_i = m_i \vec{r}_i \times (\vec{\omega} \times \vec{r}_i)$$

The total angular momentum of the body is

$$\vec{J} = \sum \vec{J}_i = \sum_i m_i \vec{r}_i \times (\vec{\omega} \times \vec{r}_i)$$

Notice that all the particles of the body rotate with the same angular velocity $\vec{\omega}$. We have

$$\vec{J} = \sum m_i [\vec{\omega}\,(\vec{r}_i \,.\, \vec{r}_i) - \vec{r}_i\,(\vec{r}_i \,.\, \vec{\omega})]$$

Since the directions of $\vec{\omega}$ and $\vec{r}_i$ are perpendicular to each other, we have $\vec{r}_i \,.\, \vec{\omega} = 0$ and therefore, we get

$$\vec{J} = \sum m_i r_i^2\, \vec{\omega} = I\, \vec{\omega}$$

where $I = \sum m_i r_i^2$ is the moment of inertia of the body about the axis of rotation. The kinetic energy of the body is

$$K = \sum \frac{1}{2}\, m_i v_i^2 = \sum \frac{1}{2}\, m_i |\, \vec{\omega} \times \vec{r}_i\,|^2$$

$$= \frac{1}{2} \sum m_i r_i^2 \omega^2 = \frac{1}{2}\, I\omega^2$$

where $I = \sum m_i r_i^2$ is the moment of inertia of the body about the axis of rotation.

3.8 Euler angles

In order to specify orientation of a rigid body in space, let us consider two systems S and S′ with common origin at O. The system S is fixed in space, whereas the system S′ is attached to a rigid body. The configuration of the body can be completely specified in the system S′. Now, orientation of the S′ relative to S represents the orientation of the body.

Euler suggested that the Cartesian coordinate measured in S can be transformed to those in S′ and vice versa by means of three successive rotations performed in a specific sequence. These rotations are known as the Euler angles.

1. The first step of the sequence of rotations is to rotate the coordinate axes (x, y, z) through an angle ϕ in the anticlockwise direction about the z-axis. Let the axes after rotation be labelled as a, b, c (Figure 3.7). The axis c is obviously same as the z-axis. From Figure 3.7, we have

$$\hat{c} = \hat{z}$$

$$\hat{a} = \cos\phi\,\hat{x} + \cos(90-\phi)\,\hat{y} = \cos\phi\,\hat{x} + \sin\phi\,\hat{y}$$

$$\hat{b} = \cos(90+\phi)\,\hat{x} + \cos\phi\,\hat{y} = -\sin\phi\,\hat{x} + \cos\phi\,\hat{y}$$

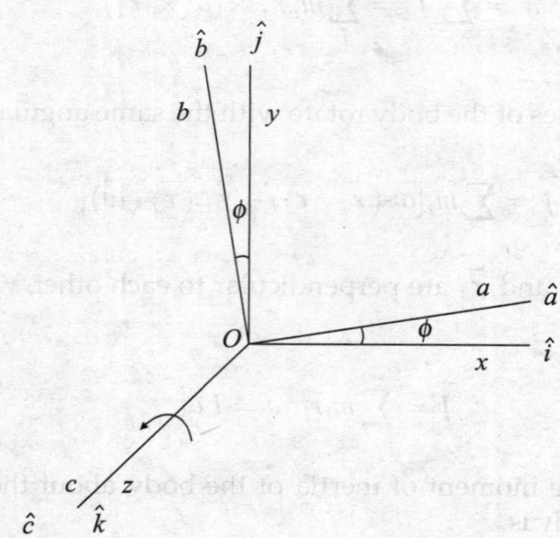

Figure 3.7: Coordinate axes (x, y, z) rotate through an angle ϕ in the anticlockwise direction about the z-axis.

These equations can be expressed as

$$\begin{bmatrix} \hat{a} \\ \hat{b} \\ \hat{c} \end{bmatrix} = \begin{bmatrix} \cos\phi & \sin\phi & 0 \\ -\sin\phi & \cos\phi & 0 \\ 0 & 0 & 1 \end{bmatrix} \begin{bmatrix} \hat{x} \\ \hat{y} \\ \hat{z} \end{bmatrix}$$

Hence, the matrix

$$\begin{bmatrix} \cos\phi & \sin\phi & 0 \\ -\sin\phi & \cos\phi & 0 \\ 0 & 0 & 1 \end{bmatrix}$$

is known as the transformation matrix. Inverse of this matrix is

$$\begin{bmatrix} \cos\phi & -\sin\phi & 0 \\ \sin\phi & \cos\phi & 0 \\ 0 & 0 & 1 \end{bmatrix}$$

2. The second step is to rotate axis a, b, c through an angle θ in the anticlockwise direction about the a-axis. Let the axes after rotation be labelled as a', b', c' (Figure 3.8). The axis a' is obviously same as the a-axis. This axis a is often referred to as the 'line of nodes'. From Figure 3.8, one can have

$$\hat{a'} = \hat{a}$$

$$\hat{b'} = \cos\theta\,\hat{b} + \cos(90-\theta)\,\hat{c} = \cos\theta\,\hat{b} + \sin\theta\,\hat{c}$$

$$\hat{c'} = \cos(90+\theta)\,\hat{b} + \cos\theta\,\hat{c} = -\sin\theta\,\hat{b} + \cos\theta\,\hat{c}$$

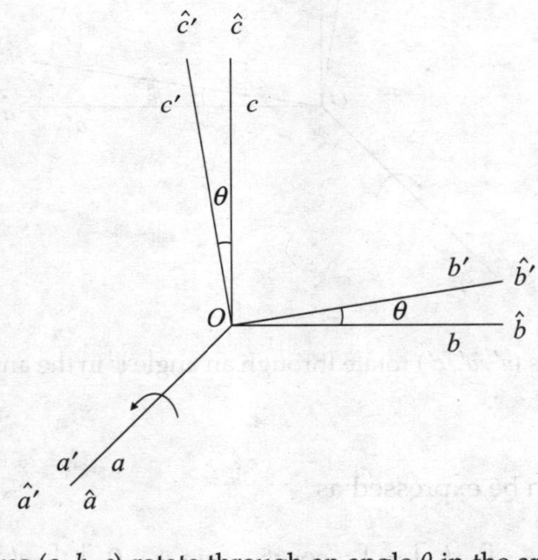

Figure 3.8: Coordinate axes (a, b, c) rotate through an angle θ in the anticlockwise direction about the a-axis.

These equations can be expressed as

$$\begin{bmatrix} \hat{a'} \\ \hat{b'} \\ \hat{c'} \end{bmatrix} = \begin{bmatrix} 1 & 0 & 0 \\ 0 & \cos\theta & \sin\theta \\ 0 & -\sin\theta & \cos\theta \end{bmatrix} \begin{bmatrix} \hat{a} \\ \hat{b} \\ \hat{c} \end{bmatrix}$$

and the inverse transformation is

$$\begin{bmatrix} \hat{a} \\ \hat{b} \\ \hat{c} \end{bmatrix} = \begin{bmatrix} 1 & 0 & 0 \\ 0 & \cos\theta & -\sin\theta \\ 0 & \sin\theta & \cos\theta \end{bmatrix} \begin{bmatrix} \hat{a}' \\ \hat{b}' \\ \hat{c}' \end{bmatrix}$$

3. The third step is to rotate the axes a', b', c' through an angle ψ in the anticlockwise direction about the c'-axis. Let the axes after rotation be labelled as x', y', z' (Figure 3.9). The axis z' is obviously same as the c'-axis. From Figure 3.9, we have

$$\hat{z}' = \hat{c}'$$

$$\hat{x}' = \cos\psi\, \hat{a}' + \cos(90 - \psi)\, \hat{b}' = \cos\psi\, \hat{a}' + \sin\psi\, \hat{b}'$$

$$\hat{y}' = \cos(90 + \psi)\, \hat{a}' + \cos\psi\, \hat{b}' = -\sin\psi\, \hat{a}' + \cos\psi\, \hat{b}'$$

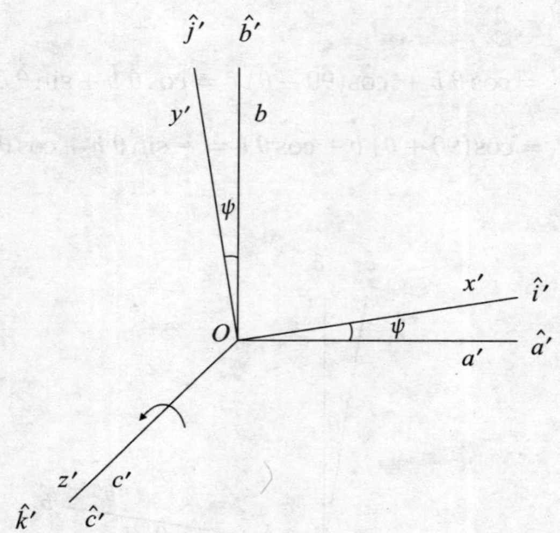

Figure 3.9: Coordinate axes (a', b', c') rotate through an angle ψ in the anticlockwise direction about the c'-axis.

These equations can be expressed as

$$\begin{bmatrix} \hat{x}' \\ \hat{y}' \\ \hat{z}' \end{bmatrix} = \begin{bmatrix} \cos\psi & \sin\psi & 0 \\ -\sin\psi & \cos\psi & 0 \\ 0 & 0 & 1 \end{bmatrix} \begin{bmatrix} \hat{a}' \\ \hat{b}' \\ \hat{c}' \end{bmatrix}$$

and the inverse transformation is

$$\begin{bmatrix} \hat{a}' \\ \hat{b}' \\ \hat{c}' \end{bmatrix} = \begin{bmatrix} \cos\psi & -\sin\psi & 0 \\ \sin\psi & \cos\psi & 0 \\ 0 & 0 & 1 \end{bmatrix} \begin{bmatrix} \hat{x}' \\ \hat{y}' \\ \hat{z}' \end{bmatrix}$$

Thus, through the sequence of three rotations (ϕ, θ, ψ), the old set of coordinates (x, y, z) can be transformed to the new set of coordinates (x', y', z'). The angles ϕ, θ, ψ are known as the Euler angles. Thus, we have

$$\begin{bmatrix} \hat{x}' \\ \hat{y}' \\ \hat{z}' \end{bmatrix} = \begin{bmatrix} \cos\psi & \sin\psi & 0 \\ -\sin\psi & \cos\psi & 0 \\ 0 & 0 & 1 \end{bmatrix} \begin{bmatrix} 1 & 0 & 0 \\ 0 & \cos\theta & \sin\theta \\ 0 & -\sin\theta & \cos\theta \end{bmatrix} \times \begin{bmatrix} \cos\phi & \sin\phi & 0 \\ -\sin\phi & \cos\phi & 0 \\ 0 & 0 & 1 \end{bmatrix} \begin{bmatrix} \hat{x} \\ \hat{y} \\ \hat{z} \end{bmatrix}$$

$$\begin{bmatrix} \hat{x}' \\ \hat{y}' \\ \hat{z}' \end{bmatrix} = \begin{bmatrix} (\cos\psi\cos\phi - \cos\theta\sin\phi\sin\psi) & (\cos\psi\sin\phi + \cos\theta\cos\phi\sin\psi) & \sin\psi\sin\theta \\ (-\sin\psi\cos\phi - \cos\theta\sin\phi\cos\psi) & (-\sin\psi\sin\phi + \cos\theta\cos\phi\cos\psi) & \cos\psi\sin\theta \\ \sin\theta\sin\phi & -\sin\theta\cos\phi & \cos\theta \end{bmatrix} \begin{bmatrix} \hat{x} \\ \hat{y} \\ \hat{z} \end{bmatrix}$$

This can be written in the form of a matrix equation

$$\hat{X}' = A\,\hat{X}$$

The inverse transformation from body coordinates to space coordinates can be obtained with the help of the equation

$$\hat{X} = A^{-1}\,\hat{X}'$$

where A^{-1} is the inverse of matrix A and can be obtained as

$$A^{-1} = \begin{bmatrix} \cos\phi & -\sin\phi & 0 \\ \sin\phi & \cos\phi & 0 \\ 0 & 0 & 1 \end{bmatrix} \begin{bmatrix} 1 & 0 & 0 \\ 0 & \cos\theta & -\sin\theta \\ 0 & \sin\theta & \cos\theta \end{bmatrix} \begin{bmatrix} \cos\psi & -\sin\psi & 0 \\ \sin\psi & \cos\psi & 0 \\ 0 & 0 & 1 \end{bmatrix}$$

$$A^{-1} = \begin{bmatrix} (\cos\psi\cos\phi - \cos\theta\sin\phi\sin\psi) & (-\sin\psi\cos\phi - \cos\theta\sin\phi\cos\psi) & \sin\phi\sin\theta \\ (\cos\psi\sin\phi + \cos\theta\cos\phi\sin\psi) & (-\sin\psi\sin\phi + \cos\theta\cos\phi\cos\psi) & -\cos\phi\sin\theta \\ \sin\theta\sin\psi & \sin\theta\cos\psi & \cos\theta \end{bmatrix}$$

Obviously, A^{-1} is transpose of A, that is $A^{-1} = \bar{A}$.

3.8.1 Angular velocity in terms of Euler angles

Suppose a body goes under three successive infinitesimal rotations ϕ, θ and ψ (i.e., Euler angles), then the three components of angular velocity $\vec{\omega}$ are $\vec{\omega}_\phi=\dot{\phi}$, $\vec{\omega}_\theta=\dot{\theta}$ and $\vec{\omega}_\psi=\dot{\psi}$, such that

$$\vec{\omega} = \vec{\omega}_\phi\ \hat{z}+ \vec{\omega}_\theta\ \hat{a}+ \vec{\omega}_\psi\ \hat{z}' = \dot{\phi}\ \hat{z}+ \dot{\theta}\ \hat{a}+ \dot{\psi}\ \hat{z}' \tag{3.44}$$

We have

$$\begin{bmatrix} \hat{a} \\ \hat{b} \\ \hat{c} \end{bmatrix} = \begin{bmatrix} 1 & 0 & 0 \\ 0 & \cos\theta & -\sin\theta \\ 0 & \sin\theta & \cos\theta \end{bmatrix} \begin{bmatrix} \cos\psi & -\sin\psi & 0 \\ \sin\psi & \cos\psi & 0 \\ 0 & 0 & 1 \end{bmatrix} \begin{bmatrix} \hat{x}' \\ \hat{y}' \\ \hat{z}' \end{bmatrix}$$

Thus,

$$\hat{a} = \cos\psi \hat{x}' - \sin\psi \hat{y}' \tag{3.45}$$

We now have

$$\begin{bmatrix} \hat{x} \\ \hat{y} \\ \hat{z} \end{bmatrix} = \begin{bmatrix} \cos\phi & -\sin\phi & 0 \\ \sin\phi & \cos\phi & 0 \\ 0 & 0 & 1 \end{bmatrix} \begin{bmatrix} 1 & 0 & 0 \\ 0 & \cos\theta & -\sin\theta \\ 0 & \sin\theta & \cos\theta \end{bmatrix} \times$$

$$\begin{bmatrix} \cos\psi & -\sin\psi & 0 \\ \sin\psi & \cos\psi & 0 \\ 0 & 0 & 1 \end{bmatrix} \begin{bmatrix} \hat{x}' \\ \hat{y}' \\ \hat{z}' \end{bmatrix}$$

It gives

$$\hat{z} = \sin\theta\ \sin\psi \hat{x}' + \sin\theta\ \cos\psi \hat{y}' + \cos\theta \hat{z}' \tag{3.46}$$

Using equations (3.45) and (3.46) in (3.44), we have

$$\begin{aligned} \vec{\omega} &= \dot{\phi}\ (\sin\theta\ \sin\psi \hat{x}'+ \sin\theta\ \cos\psi \hat{y}'+ \cos\theta \hat{z}')+ \dot{\theta}\ (\cos\psi \hat{x}' - \sin\psi \hat{y}')+ \dot{\psi} \hat{z}' \\ &= \hat{x}' (\dot{\phi}\sin\theta\ \sin\psi+ \dot{\theta}\cos\psi) + \hat{y}'(\dot{\phi}\sin\theta\ \cos\psi- \dot{\theta}\sin\psi) + \hat{z}'(\dot{\phi}\cos\theta+ \dot{\psi}) \\ &= \omega_{x'}\hat{x}' + \omega_{y'}\hat{y}' + \omega_{z'}\hat{z}' \end{aligned} \tag{3.47}$$

where

$$\omega_{x'} = \dot{\phi}\sin\theta\ \sin\psi+ \dot{\theta}\cos\psi \qquad \omega_{y'} = \dot{\phi}\sin\theta\ \cos\psi- \dot{\theta}\sin\psi$$

$$\omega_{z'} = \dot{\phi}\cos\theta+ \dot{\psi} \tag{3.48}$$

3.8.2 Kinetic energy of rotating body

If I_1, I_2, I_3 are moments of inertia about the axes x', y', z', then the kinetic energy is

$$\begin{aligned} T = {} & \frac{I_1}{2}(\dot{\phi}\sin\theta\ \sin\psi+ \dot{\theta}\cos\psi)^2 + \frac{I_2}{2}(\dot{\phi}\sin\theta\ \cos\psi- \dot{\theta}\sin\psi)^2 \\ & + \frac{I_3}{2}(\dot{\phi}\cos\theta+ \dot{\psi})^2 \end{aligned} \tag{3.49}$$

3.9 Moments of inertia

Let us consider a body consisting of a number of particles rotating about an axis. If m_i and $\vec{r}_i$ are, respectively, mass and position vector of ith particle of the body, the moment of inertia of the body about the axis of rotation is

$$I = \sum_i m_i r_i^2$$

Exercise 3.14: Two particles of masses 1 and 2 units are placed at the points having coordinates (2,3) and (4,5), respectively. Calculate the moment of inertia of the system of particles about x-axis and about y-axis.

Solution: The moment of inertia of the system of particles about x-axis is

$$I_x = 1(3^2) + 2(5^2) = 59 \text{ units}$$

The moment of inertia of the system of particles about y-axis is

$$I_y = 1(2^2) + 2(4^2) = 36 \text{ units}$$

3.9.1 Theorem of parallel axis

Suppose we have a body of mass M whose centre of mass is at a point C. Let us consider two parallel axes separated by a distance d and one of the axis passing through the centre of mass of the body. Let I_c be the moment of inertia of the body about the axis passing through the centre of mass of the body. The theorem of parallel axis states that the moment of inertia I of the body about the second axis is

$$I = I_c + Md^2$$

Proof: Proof of the theorem can be obtained in the following manner. Consider a body of mass M having large number of particles. Suppose two parallel axes separated by a distance d are intersecting a plane at two points O and C. The axis passing through the centre of mass intersects the plane at the point C.

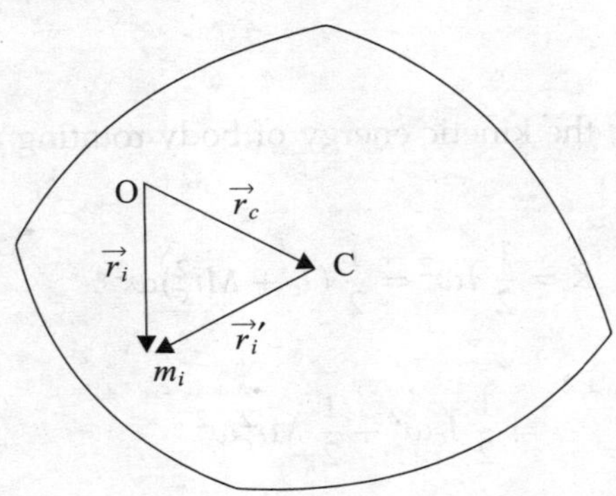

Suppose $\vec{r}_i$ is the position vector of mass m_i relative to the origin at O. The moment of inertia of the body about the axis passing through O is

$$I = \sum_i m_i r_i^2 = \sum_i m_i \vec{r}_i^{\,2} \tag{3.50}$$

If $\vec{r}_c$ is the position vector of point C relative to the point O and $\vec{r}\,'_i$ is the position vector of mass m_i relative to C, we have

$$\vec{r}_i = \vec{r}_c + \vec{r}\,'_i \tag{3.51}$$

Using equation (3.51) in (3.50), we get

$$I = \sum_i m_i (\vec{r}_c + \vec{r}\,'_i)^2 = \sum_i m_i (r_c^2 + \vec{r}_c \cdot \vec{r}\,'_i + r_i'^2)$$

$$= r_c^2 \sum_i m_i + \vec{r}_c \cdot \sum_i m_i \vec{r}\,'_i + \sum_i m_i r_i'^2 \tag{3.52}$$

When m_i is mass of ith particle of the body whose position vector relative to the centre of mass is $\vec{r}\,'_i$, then we have

$$\sum m_i \vec{r}\,'_i = 0 \tag{3.53}$$

Further, the moment of inertia of the body about the axis passing through C is

$$I_c = \sum_i m_i r_i'^2 \tag{3.54}$$

We have $M = \sum_i m_i$. On using equations (3.53) and (3.54) in (3.52), we get

$$I = I_c + M r_c^2 \tag{3.55}$$

If the separation between two axes passing through O and C is d, we have $r_c = d$ and therefore, we have

$$I = I_c + M d^2$$

Kinetic energy

In view of equation (3.55), the kinetic energy of body rotating about the axis of rotation passing through O is

$$K = \frac{1}{2} I\omega^2 = \frac{1}{2} (I_c + M r_c^2)\omega^2$$

$$= \frac{1}{2} I_c\omega^2 + \frac{1}{2} M r_c^2 \omega^2$$

The first part of kinetic energy is due to rotation of body with angular velocity ω about the axis of rotation passing through the centre of mass. The second part is due to the translational (rotational) motion of the centre of mass about the axis of rotation.

Angular momentum

Further, in view of equation (3.55), the angular momentum of body rotating about the axis passing through O is

$$J = I\omega = (I_c + Mr_c^2)\omega = I_c\omega + Mr_c^2\omega$$

The first part of angular momentum is due to rotation of body with angular velocity ω about the axis of rotation passing through the centre of mass. The second part is due to the translational (rotational) motion of the centre of mass about the axis of rotation.

Exercise 3.15: In a plane lamina of mass 10 kg, there are three points A, B, C forming a triangle, as shown below. Point C is the centre of mass of the lamina. Three parallel axes, perpendicular to the lamina are passing through A, B and C. If the moment of inertia of the body about the axis passing through point A is 500 kg-m^2. Calculate the moment of inertia of the body about the axis passing through point B.

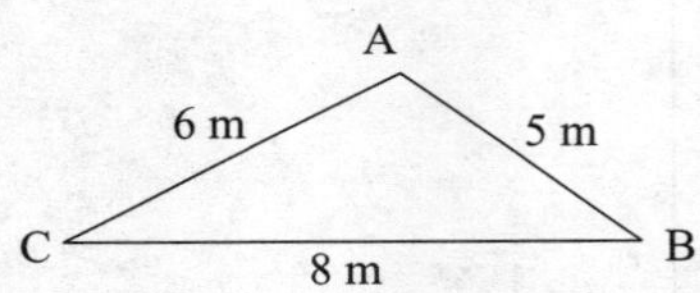

Solution: Total mass of the body M = 10 kg and moment of inertia of the body about the axis passing through A is I_A = 500 kg-m^2. Let I_C be the moment of inertia of the body about the axis passing through C. From the theorem of parallel axis we have

$$I_A = I_c + M6^2$$

Using the values, we have

$$500 = I_c + 10 \times (6^2) \qquad \text{and therefore} \qquad I_c = 140 \text{ kg-m}^2$$

The moment of inertia about the axis passing through B is

$$I_B = I_c + M8^2$$

Using the values, we have

$$I_B = 140 + 10 \times (8^2) = 780 \text{ kg-m}^2$$

Note: One should be careful not to apply the theorem of parallel axis directly between the axes passing through the points A and B. It should be applied through the centre of mass C.

3.9.2 Theorem of perpendicular axis

The theorem of perpendicular axis states that the moment of inertia of a plane lamina about an axis perpendicular to its plane is equal to the sum of the moments of inertia about any two perpendicular axes in the lamina and intersecting the first (perpendicular) axis. If in a plane lamina, the two perpendicular axes are denoted by x-axis and y-axis, the axis perpendicular to the plane lamina is z-axis. If I_x, I_y, I_z are moments of inertia of the lamina about x-axis, y-axis and z-axis, respectively, the theorem of perpendicular axis states that

$$I_z = I_x + I_y$$

Proof: Proof of the theorem can be obtained in the following manner. Consider a plane lamina having a large number of particles. Suppose m_i is the mass of ith particle situated at a point (x_i, y_i). The moment of inertia of the lamina about x-axis is

$$I_x = \sum_i m_i y_i^2$$

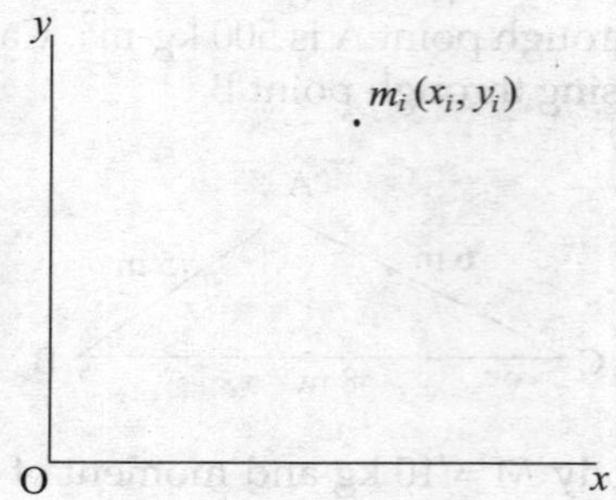

and about y-axis is

$$I_y = \sum_i m_i x_i^2$$

The moment of inertia of the lamina about z-axis is

$$I_z = \sum_i m_i r_i^2$$

We have $r_i^2 = x_i^2 + y_i^2$ and therefore, we have

$$I_z = \sum_i m_i(x_i^2 + y_i^2) = \sum_i m_i x_i^2 + \sum_i m_i y_i^2 = I_y + I_x$$

Note: It may be noticed that this theorem is applicable for a plane lamina only.

Exercise 3.16: In a plane, two particles A and B of masses 2 kg and 3 kg are placed at the points with coordinates $(-2,3)$ and $(4,-5)$, respectively. Calculate the moments of inertia of the system of two particles about x-axis, y-axis and z-axis. Check if the theorem of perpendicular axis is satisfied or not.

Solution: The moment of inertia of the system of two particles about x-axis is

$$I_x = 2(3)^2 + 3(-5)^2 = 93 \text{ units}$$

The moment of inertia of the system about y-axis is

$$I_y = 2(-2)^2 + 3(4)^2 = 56 \text{ units}$$

The distance of particle A from z-axis is

$$d_A = \sqrt{(-2)^2 + (3)^2} = \sqrt{13}$$

The distance of particle B from z-axis is

$$d_B = \sqrt{(4)^2 + (-5)^2} = \sqrt{41}$$

The moment of inertia of the system about z-axis is

$$I_z = 2(\sqrt{13})^2 + 3(\sqrt{41})^2 = 149 \text{ units}$$

Since $I_z = I_x + I_y$, the theorem of perpendicular axis is satisfied.

3.9.3 Moment of inertia of a rectangular lamina

Let us consider a uniform rectangular lamina of length a, width b and mass m. Lamina is a shape whose thickness is negligibly small as compared to the length and width. We want to calculate moment of inertia of the lamina about y axis (parallel to the width of the lamina), passing through the centre of mass and lying in the plane of the lamina. The mass per unit length of the lamina is

$$\sigma = \frac{m}{a}$$

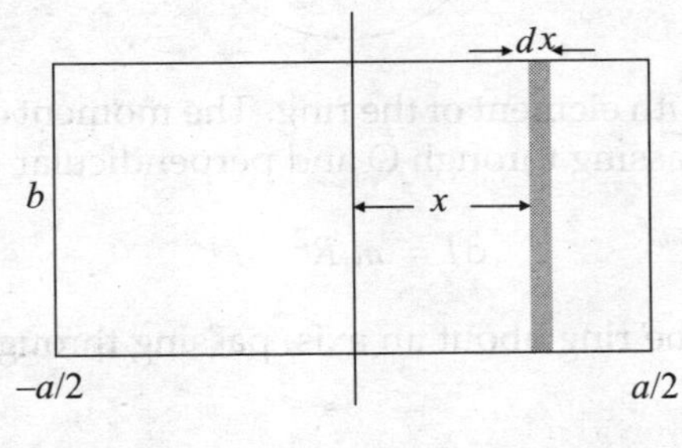

Let us consider a strip of thickness dx at a distance x from the y-axis. The mass of the strip is

$$dm = \sigma\, dx = \frac{m}{a}\, dx$$

The moment of inertia of the strip about the y-axis is

$$dI_y = dm\, x^2 = \frac{m}{a} x^2 dx$$

Thus, the moment of inertia of the lamina about the y-axis is

$$I_y = \int_{-a/2}^{a/2} \frac{m}{a} x^2 dx = \frac{m}{a}\left[\frac{x^3}{3}\right]_{-a/2}^{a/2} = \frac{ma^2}{12}$$

Similarly, we can find out the moment of inertia of the lamina about x axis, passing through the centre of mass and lying in the plane of the lamina as

$$I_x = \frac{mb^2}{12}$$

By applying the theorem of the perpendicular axis, the moment of inertia of the lamina about z-axis and passing through the centre of mass is

$$I_z = I_x + I_y = \frac{mb^2}{12} + \frac{ma^2}{12} = \frac{m(a^2 + b^2)}{12}$$

3.9.4 Moment of inertia of a ring

Let us consider a ring of radius R and mass m. We want to calculate moment of inertia of the ring about an axis, passing through the centre O of the ring and perpendicular to the plane of the ring (plane of the paper). Let us assume that the ring is made up of a large number of elements such that

$$m = \sum_i m_i$$

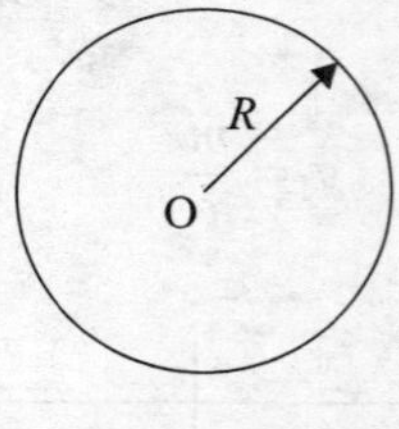

where m_i is the mass of the ith element of the ring. The moment of inertia of the ith element of the ring about an axis, passing through O and perpendicular to the plane of the ring is

$$\mathrm{d}I = m_i R^2$$

The moment of inertia of the ring about an axis, passing through O and perpendicular to the plane of the ring is

$$I = \sum_i m_i R^2 = mR^2$$

Moment of inertia of ring about an axis passing through O and lying in the plane of the ring

Suppose I' is the moment of inertia of the ring about an axis passing through O and lying in the plane of the ring. We can consider two such axes which are mutually perpendicular to each other, as shown in the following figure. From the theorem of perpendicular axis, we have

$$I = I' + I' \qquad \text{or} \qquad I' = \frac{I}{2} = \frac{mR^2}{2}$$

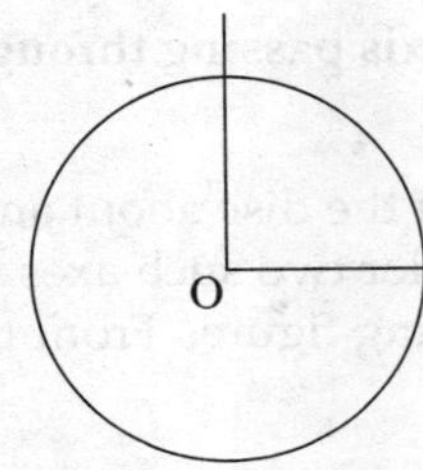

3.9.5 Moment of inertia of a disc

Let us consider a uniform circular disc of mass M and radius R. We want to calculate moment of inertia of the disc about an axis passing through the centre O and perpendicular to the plane of the disc. The mass per unit area is

$$\sigma = \frac{M}{\pi R^2}$$

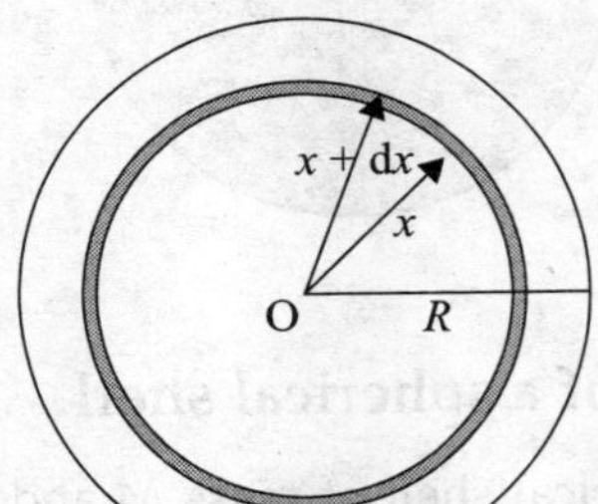

Consider a ring of radius x and thickness dx. The area of the ring is

$$A = \pi(x + \mathrm{d}x)^2 - \pi x^2 = 2\pi x\,\mathrm{d}x$$

Here, smaller term is neglected. The mass of the ring is

$$m = \sigma A = \frac{2M}{R^2}\,x\,\mathrm{d}x$$

The moment of inertia of the ring about an axis passing through the centre O and perpendicular to the plane of the disc is

$$\mathrm{d}I = mx^2 = \frac{2M}{R^2}\,x^3\,\mathrm{d}x$$

The moment of inertia of the disc about an axis passing through the centre O and perpendicular to the plane of the disc is

$$I = \frac{2M}{R^2}\int_0^R x^3\,\mathrm{d}x = \frac{2M}{R^2}\left[\frac{x^4}{4}\right]_0^R = \frac{1}{2}\,MR^2$$

Moment of inertia of disc about an axis passing through O and lying in the plane of the disc

Suppose I' is the moment of inertia of the disc about an axis passing through O and lying in the plane of the disc. We can consider two such axes which are mutually perpendicular to each other, as shown in the following figure. From the theorem of perpendicular axis, we have

$$I = I' + I' \qquad \text{or} \qquad I' = \frac{I}{2} = \frac{mR^2}{4}$$

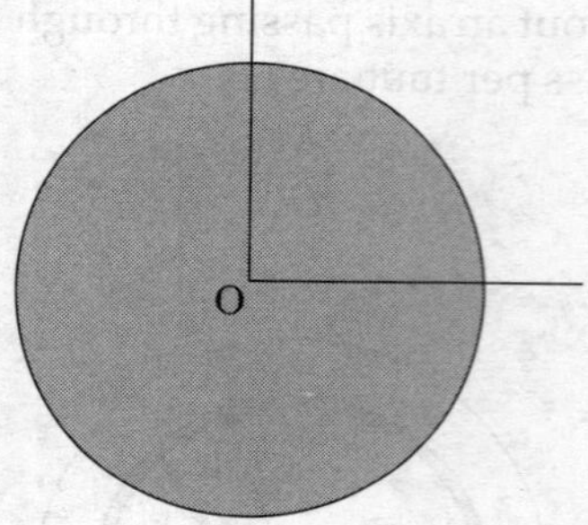

3.9.6 Moment of inertia of a spherical shell

Let us consider a uniform spherical shell of mass M and radius R. We want to calculate moment of inertia of the shell about an axis PP′ passing through the centre O of the shell. The mass per unit surface area of the shell is

$$\sigma = \frac{M}{4\pi R^2}$$

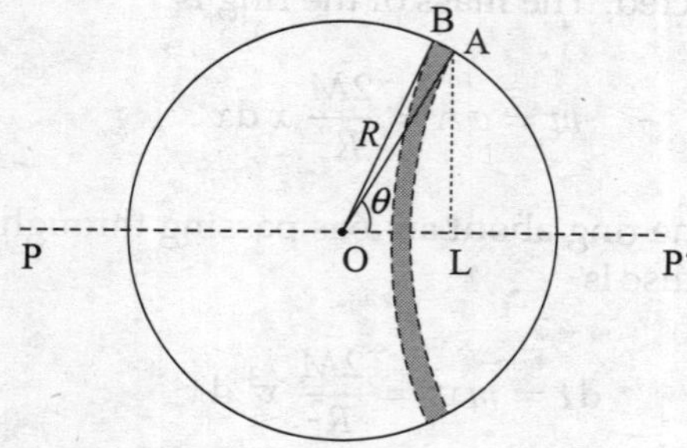

Let us consider a ring on the shell having angular width $d\theta$ so that $AB = R\,d\theta$. The radius of the ring is $AL = R\sin\theta$ and therefore the circumference of the ring is $2\pi R\sin\theta$. The area of the ring is

$$(2\pi R\sin\theta)(R\,d\theta) = 2\pi R^2 \sin\theta\, d\theta$$

and mass of the ring is

$$m_r = 2\pi R^2 \sin\theta \, \mathrm{d}\theta \, \sigma = \frac{M \sin\theta \, \mathrm{d}\theta}{2}$$

Here, we used the value of the surface density σ. The ring may be considered as consisting of a large number of small elements so that $m_r = \sum m_i$, where m_i is the mass of the ith element. The moment of inertia of the ith element about PP′ is $m_i(R\sin\theta)^2$. The moment of inertia of the ring about PP′ is

$$\mathrm{d}I = \sum_i m_i (R\sin\theta)^2 = m_r R^2 \sin^2\theta$$

Using the expression for m_r, we have

$$\mathrm{d}I = \frac{MR^2 \sin^3\theta \, \mathrm{d}\theta}{2}$$

The moment of inertia of the shell about PP′ is

$$I = \frac{MR^2}{2} \int_0^{\pi} \sin^3\theta \, \mathrm{d}\theta = \frac{MR^2}{2} \int_0^{\pi} \sin\theta (1 - \cos^2\theta) \, \mathrm{d}\theta$$

Putting $x = \cos\theta$ so that $\mathrm{d}x = -\sin\theta \, \mathrm{d}\theta$, we have

$$I = \frac{MR^2}{2} \int_{-1}^{1} (1 - x^2) \, \mathrm{d}x = \frac{2MR^2}{3}$$

3.9.7 Moment of inertia of a solid sphere

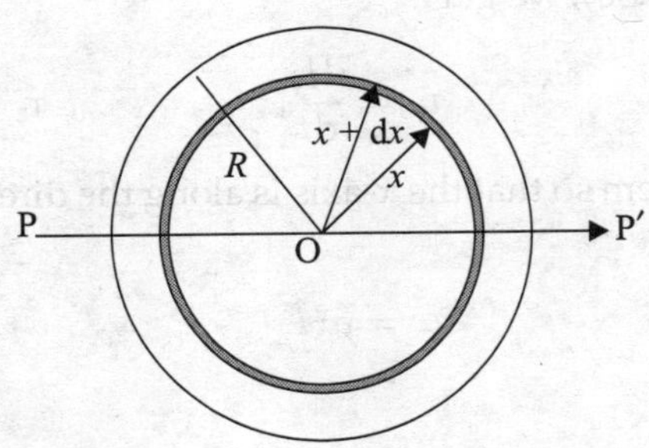

Figure 3.10: Spherical shell of radius x and thickness $x + \mathrm{d}x$.

Let us consider a solid sphere of mass M and radius R. We want to calculate the moment of inertia of the solid sphere about an axis PP′ passing through the centre O of the sphere. The mass per unit volume is

$$\rho = \frac{M}{4\pi R^3/3}$$

Let us consider a spherical shell of radius x and thickness $\mathrm{d}x$ (Figure 3.10). The volume of the shell is

$$V = \frac{4}{3}\,\pi(x+\mathrm{d}x)^3 - \frac{4}{3}\,\pi x^3 = 4\pi x^2\,\mathrm{d}x$$

The mass of the shell is

$$m = \rho V = \frac{3M}{R^3}\,x^2\,\mathrm{d}x$$

Using the knowledge of moment of inertia about an axis PP′ passing through O, we have

$$\mathrm{d}I = \frac{2mx^2}{3} = \frac{2M}{R^3}\,x^4\,\mathrm{d}x$$

Here, we used the expression for m. The moment of inertia of solid spherical shell about an axis PP′ passing through O is

$$I = \frac{2M}{R^3}\int_0^R x^4\,\mathrm{d}x = \frac{2M}{R^3}\left[\frac{x^5}{5}\right]_0^R = \frac{2}{5}\,MR^2$$

3.10 Rotation about a fixed axis

For rotational motion, the equation of motion is

$$\vec{\tau} = \frac{\mathrm{d}\vec{J}}{\mathrm{d}t} \tag{3.56}$$

The torque $\vec{\tau}$ and angular momentum $\vec{J}$ may be expressed as

$$\vec{\tau} = \tau_x\,\hat{\imath} + \tau_y\,\hat{\jmath} + \tau_z\,\hat{k} \qquad \text{and} \qquad \vec{J} = J_x\,\hat{\imath} + J_y\,\hat{\jmath} + J_z\,\hat{k} \tag{3.57}$$

Using equations (3.57) in (3.56), we get

$$\tau_x = \frac{\mathrm{d}J_x}{\mathrm{d}t} \qquad \tau_y = \frac{\mathrm{d}J_y}{\mathrm{d}t} \qquad \tau_z = \frac{\mathrm{d}J_z}{\mathrm{d}t}$$

We take the coordinate system so that the x-axis is along the direction of the axis of rotation. Thus, we have

$$\vec{\omega} = \omega\,\hat{\imath}$$

Therefore, we have

$$\vec{J}\,.\,\vec{\omega} = J_x\omega \qquad \text{or} \qquad J_x = \frac{\vec{J}\,.\,\vec{\omega}}{\omega} \tag{3.58}$$

Let us consider a body consisting of a large number of particles rotating with angular velocity $\vec{\omega}$ about x-axis. If m_i be the mass of ith particle situated at a position vector $\vec{r}_i$, then total momentum of the body is

$$\vec{J} = \sum_i \vec{r}_i \times m_i(\vec{\omega} \times \vec{r}_i) \tag{3.59}$$

Using equation (3.59) in (3.58), we get

$$J_x = \frac{1}{\omega}\left[\sum_i \vec{r}_i \times m_i(\vec{\omega} \times \vec{r}_i)\right].\vec{\omega}$$

Using the relation $\vec{A}\,.\,\vec{B} = \vec{B}\,.\,\vec{A}$, we have

$$J_x = \frac{1}{\omega}\sum_i m_i\,\vec{\omega}\,.[\vec{r}_i \times (\vec{\omega} \times \vec{r}_i)]$$

Using the relation $\vec{A}\,.(\vec{B} \times \vec{C}) = (\vec{A} \times \vec{B}).\,\vec{C}$

$$J_x = \frac{1}{\omega}\sum_i m_i(\vec{\omega} \times \vec{r}_i).(\vec{\omega} \times \vec{r}_i) = \frac{1}{\omega}\sum_i m_i|\,\vec{\omega} \times \vec{r}_i\,|^2$$

$$= \frac{1}{\omega}\sum_i m_i(\omega r_i \sin\theta_i)^2 = \omega\sum_i m_i R_i^2 = I_x\omega$$

where θ_i is the angle as shown in the figure below, and $I_x = \sum_i m_i R_i^2$ is the moment of inertia of body about x-axis. Now, we have

$$\vec{\tau}\,.\,\vec{\omega} = \tau_x\,\omega \qquad \text{or} \qquad \tau_x = \frac{\vec{\tau}\,.\,\vec{\omega}}{\omega}$$

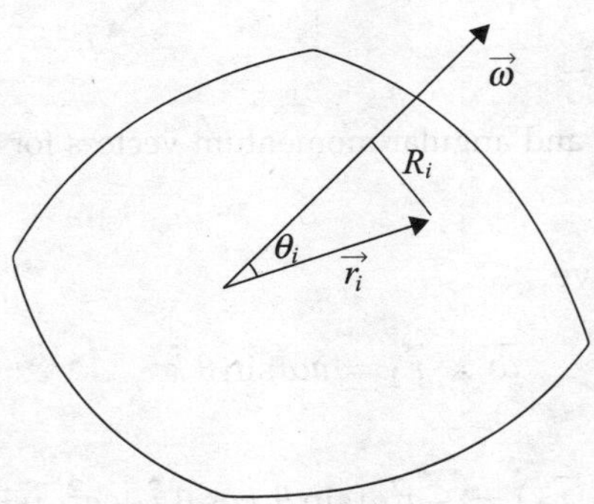

The kinetic energy for x-axis is

$$K_x = \frac{1}{2}\,I_x\omega^2$$

When the angular momentum $\vec{J}$ does not coincide with the axis of rotation, its time derivative $\mathrm{d}\vec{J}/\mathrm{d}t$ can involve a changing direction of the vector $\vec{J}$ as well as a varying magnitude. A rotation about a fixed axis implies a circular motion of all the features of the body.

As a simple case, let us consider a light rod with two equal masses at its ends and rotating about a fixed axis passing through the centre of mass and oriented at an angle θ from the rod. At an instant, the rod is in the xy-plane, as shown in Figure 3.11. The rod is of length $2a$ and the angular velocity is represented by $\vec{\omega}$ which lies along the x-axis. The total angular momentum of the rod is

$$\vec{J} = \vec{r}_1 \times m(\vec{\omega} \times \vec{r}_1) + \vec{r}_2 \times m(\vec{\omega} \times \vec{r}_2) \tag{3.60}$$

where $\vec{r}_1$ and $\vec{r}_2$ are position vectors of masses about the centre of mass. We can write as following:

$$\vec{r}_1 = a\cos\theta\,\hat{\imath} + a\sin\theta\,\hat{\jmath}$$

$$\vec{r}_2 = -a\cos\theta\,\hat{\imath} - a\sin\theta\,\hat{\jmath}$$

$$\vec{\omega} = \omega\,\hat{\imath}$$

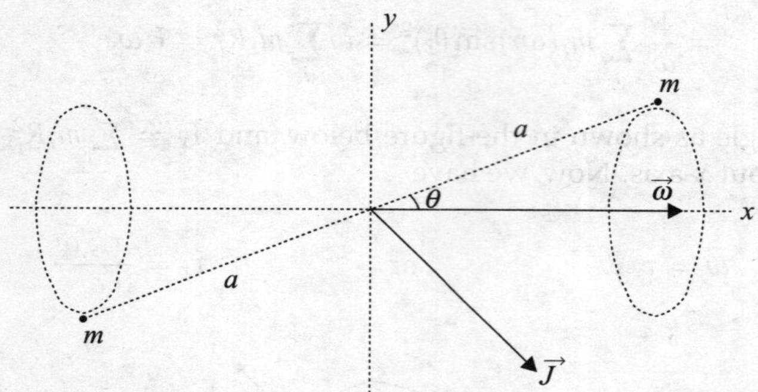

Figure 3.11: Angular velocity and angular momentum vectors for a light rod with masses at the ends.

For these expressions, we have

$$\vec{\omega} \times \vec{r}_1 = a\omega\sin\theta\,\hat{k}$$

and

$$\vec{r}_1 \times (\vec{\omega} \times \vec{r}_1) = -a^2\omega\sin\theta\,\cos\theta\,\hat{\jmath} + a^2\omega\sin^2\theta\,\hat{\imath} \tag{3.61}$$

Further,

$$\vec{\omega} \times \vec{r}_2 = -a\omega\sin\theta\,\hat{k}$$

and

$$\vec{r}_2 \times (\vec{\omega} \times \vec{r}_2) = -a^2\omega\sin\theta\,\cos\theta\,\hat{\jmath} + a^2\omega\sin^2\theta\,\hat{\imath} \tag{3.62}$$

Using equations (3.61) and (3.62) in (3.60), we get

$$\vec{J} = 2ma^2\omega\sin\theta(\sin\theta\,\hat{\imath} - \cos\theta\,\hat{\jmath})$$

This vector is perpendicular to the rod, in the orientation shown in the figure. It rotates about the x-axis, always keeping a same relationship to the rod and to the axis of rotation.

Since $\vec{J}$ rotates, its time derivative is not zero. In this case, we have[6]

$$\frac{d\vec{J}}{dt} = \vec{\omega} \times \vec{J}$$

Now, we have

$$\vec{\omega} \times \vec{J} = -2ma^2\omega^2 \sin\theta \, \cos\theta \, \hat{k}$$

and, we get

$$\frac{d\vec{J}}{dt} = -2ma^2\omega^2 \sin\theta \, \cos\theta \, \hat{k}$$

where $\hat{k}$ is the direction perpendicular to the xy-plane. It shows that the angular momentum will be constant when either $\theta = 0$ or $\theta = 90°$. In fact, the torque is

$$\vec{\tau} = \frac{d\vec{J}}{dt} = -2ma^2\omega^2 \sin\theta \, \cos\theta \, \hat{k}$$

When the torque $\vec{\tau}$ acting on the system is zero, we have either $\theta = 0$ or $\theta = 90°$.

3.11 Moments and products of inertia

Consider a body consisting of a large number of particles rotating with angular velocity $\vec{\omega}$ about an axis. If m_i be the mass of ith particle situated at a position vector $\vec{r}_i$, then total momentum of the body is

$$\vec{J} = \sum \vec{r}_i \times m_i(\vec{\omega} \times \vec{r}_i) \tag{3.63}$$

On expressing $\vec{r}_i$ and $\vec{\omega}$ in Cartesian coordinates as

$$\vec{r}_i = x_i\,\hat{\imath} + y_i\,\hat{\jmath} + z_i\,\hat{k}$$

$$\vec{\omega} = \omega_x\,\hat{\imath} + \omega_y\,\hat{\jmath} + \omega_z\,\hat{k}$$

we have

$$\vec{\omega} \times \vec{r}_i = \begin{vmatrix} \hat{\imath} & \hat{\jmath} & \hat{k} \\ \omega_x & \omega_y & \omega_z \\ x_i & y_i & z_i \end{vmatrix}$$

[6] Velocity can be expressed as

$$\vec{v} = \frac{d\vec{r}}{dt} = \vec{\omega} \times \vec{r}$$

Thus, the time-derivative operator can be expressed in case of rotational motion as

$$\frac{d}{dt} = \vec{\omega} \times$$

$$= (\omega_y z_i - \omega_z y_i)\hat{\imath} + (\omega_z x_i - \omega_x z_i)\hat{\jmath} + (\omega_x y_i - \omega_y x_i)\hat{k}$$

and

$$\vec{r}_i \times (\vec{\omega} \times \vec{r}_i) = \begin{vmatrix} \hat{\imath} & \hat{\jmath} & \hat{k} \\ x_i & y_i & z_i \\ (\omega_y z_i - \omega_z y_i) & (\omega_z x_i - \omega_x z_i) & (\omega_x y_i - \omega_y x_i) \end{vmatrix}$$

$$= [y_i(\omega_x y_i - \omega_y x_i) - z_i(\omega_z x_i - \omega_x z_i)]\hat{\imath} + [z_i(\omega_y z_i - \omega_z y_i)$$

$$- x_i(\omega_x y_i - \omega_y x_i)]\hat{\jmath} + [x_i(\omega_z x_i - \omega_x z_i) - y_i(\omega_y z_i - \omega_z y_i)]\hat{k} \tag{3.64}$$

Using equation (3.64) in (3.63) and rearranging, we get

$$\vec{J} = \sum_i m_i\{[\omega_x(y_i^2 + z_i^2) - \omega_y x_i y_i - \omega_z x_i z_i]\hat{\imath} + [\omega_y(z_i^2 + x_i^2) - \omega_z y_i z_i$$

$$- \omega_x y_i x_i]\hat{\jmath} + [\omega_z(x_i^2 + y_i^2) - \omega_x z_i x_i - \omega_y z_i y_i]\hat{k}\} \tag{3.65}$$

Considering

$$\vec{J} = J_x\,\hat{\imath} + J_y\,\hat{\jmath} + J_z\,\hat{k} \tag{3.66}$$

and comparing coefficients of $\hat{\imath}$, $\hat{\jmath}$ and $\hat{k}$ in equations (3.65) and (3.66), we get

$$J_x = \omega_x \sum_i m_i(y_i^2 + z_i^2) - \omega_y \sum_i m_i x_i y_i - \omega_z \sum_i m_i x_i z_i$$

$$J_y = \omega_y \sum_i m_i(z_i^2 + x_i^2) - \omega_z \sum_i m_i y_i z_i - \omega_x \sum_i m_i y_i x_i$$

$$J_z = \omega_z \sum_i m_i(x_i^2 + y_i^2) - \omega_x \sum_i m_i z_i x_i - \omega_y \sum_i m_i z_i y_i$$

These expressions can be expressed as

$$J_x = I_{xx}\omega_x + I_{xy}\omega_y + I_{xz}\omega_z \tag{3.67}$$

$$J_y = I_{yx}\omega_x + I_{yy}\omega_y + I_{yz}\omega_z \tag{3.68}$$

$$J_z = I_{zx}\omega_x + I_{zy}\omega_y + I_{zz}\omega_z \tag{3.69}$$

where

$$I_{xx} = \sum_i m_i(y_i^2 + z_i^2) \qquad I_{xy} = -\sum_i m_i x_i y_i \qquad I_{xz} = -\sum_i m_i x_i z_i$$

$$I_{yx} = -\sum_i m_i y_i x_i \qquad I_{yy} = \sum_i m_i(z_i^2 + x_i^2) \qquad I_{yz} = -\sum_i m_i y_i z_i$$

$$I_{zx} = -\sum_i m_i z_i x_i \qquad I_{zy} = -\sum_i m_i z_i y_i \qquad I_{zz} = \sum_i m_i(x_i^2 + y_i^2)$$

Here, I_{xx}, I_{yy}, I_{zz} are moments of inertia about x, y, z axes, respectively. I_{xy}, I_{xz}, I_{yx}, I_{yz}, I_{zx}, I_{zy} are products of inertia. Equations (3.67), (3.68), (3.69) can be rearranged as

$$\begin{bmatrix} J_x \\ J_y \\ J_z \end{bmatrix} = \begin{bmatrix} I_{xx} & I_{xy} & I_{xz} \\ I_{yx} & I_{yy} & I_{yz} \\ I_{zx} & I_{zy} & I_{zz} \end{bmatrix} \begin{bmatrix} \omega_x \\ \omega_y \\ \omega_z \end{bmatrix}$$

The square matrix on the right side of this equation is often known as the inertia tensor. This representation shows that the diagonal members are the moments of inertia about the respective axis. For example, I_{xx} is moment of inertia about x-axis, as $y_i^2 + z_i^2$ is simply the square of the distance of ith particle from x-axis. The off-diagonal elements are known as the products of inertia and they are symmetrical in pairs, for example $I_{xy} = I_{yx}$.

The aforementioned expressions for moments and products of inertia are valid when the masses are discrete. For solid bodies, the distribution of mass is continuous. Suppose ρ be the density of the body. The density ρ may be a function of (x, y, z) or constant. For continuous variation of density, the m_i is replaced by $\rho\, dx\, dy\, dz$, the summation sign changes to integrals and the corresponding expressions are as following:

$$I_{xx} = \iiint \rho(y^2 + z^2)\, dx\, dy\, dz$$

$$I_{xy} = -\iiint \rho xy\, dx\, dy\, dz$$

$$I_{xz} = -\iiint \rho xz\, dx\, dy\, dz$$

$$I_{yx} = -\iiint \rho yx\, dx\, dy\, dz$$

$$I_{yy} = \iiint \rho(z^2 + x^2)\, dx\, dy\, dz$$

$$I_{yz} = -\iiint \rho yz\, dx\, dy\, dz$$

$$I_{zx} = -\iiint \rho zx\, dx\, dy\, dz$$

$$I_{zy} = -\iiint \rho zy\, dx\, dy\, dz$$

$$I_{zz} = \iiint \rho(x^2 + y^2)\, dx\, dy\, dz$$

Exercise 3.17: Consider a uniform cube of mass m, with each side a. Define a coordinate system with the origin at one corner and three axes along the three adjacent edges of the cube. Calculate the inertia tensor for this cube with respect to the coordinate system.

Solution: A cube having each side a is as shown in the following figure.

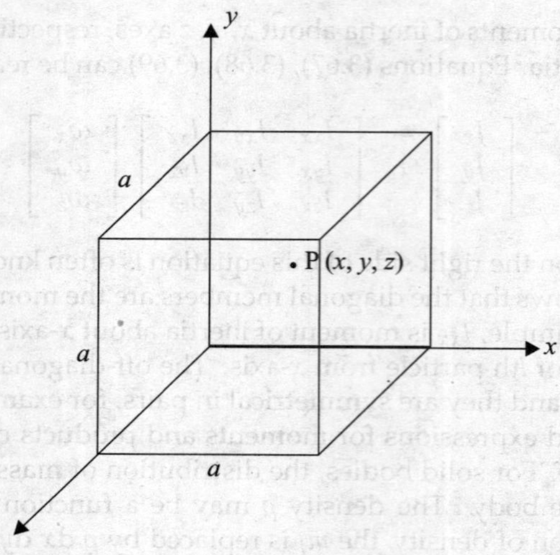

The volume of the cube is a^3. Thus, the constant density ρ is

$$\rho = \frac{m}{a^3}$$

We have

$$I_{xx} = \iiint \rho(y^2 + z^2)\, dx\, dy\, dz$$

$$= \rho \int_0^a dx \int_0^a dz \int_0^a y^2\, dy + \rho \int_0^a dx \int_0^a dy \int_0^a z^2\, dz$$

$$= \frac{\rho a^5}{3} + \frac{\rho a^5}{3} = \frac{2\rho a^5}{3} = \frac{2ma^2}{3}$$

$$I_{xy} = -\iiint \rho xy\, dx\, dy\, dz$$

$$= -\rho \int_0^a dz \int_0^a x\, dx \int_0^a y\, dy = -\frac{\rho a^5}{4} = -\frac{ma^2}{4}$$

$$I_{xz} = -\iiint \rho xz\, dx\, dy\, dz$$

$$= -\rho \int_0^a dy \int_0^a x\, dx \int_0^a z\, dz = \frac{\rho a^5}{4} = -\frac{ma^2}{4}$$

$$I_{yx} = -\iiint \rho yx\, \mathrm{d}x\, \mathrm{d}y\, \mathrm{d}z$$

$$= -\rho \int_0^a \mathrm{d}z \int_0^a x\, \mathrm{d}x \int_0^a y\, \mathrm{d}y = -\frac{\rho a^5}{4} = -\frac{ma^2}{4}$$

$$I_{yy} = \iiint \rho(z^2 + x^2)\, \mathrm{d}x\, \mathrm{d}y\, \mathrm{d}z$$

$$= \rho \int_0^a \mathrm{d}x \int_0^a \mathrm{d}y \int_0^a z^2\, \mathrm{d}z + \rho \int_0^a \mathrm{d}y \int_0^a \mathrm{d}z \int_0^a x^2\, \mathrm{d}x$$

$$= \frac{\rho a^5}{3} + \frac{\rho a^5}{3} = \frac{2\rho a^5}{3} = \frac{2ma^2}{3}$$

$$I_{yz} = -\iiint \rho yz\, \mathrm{d}x\, \mathrm{d}y\, \mathrm{d}z$$

$$= -\rho \int_0^a \mathrm{d}x \int_0^a y\, \mathrm{d}y \int_0^a z\, \mathrm{d}z = -\frac{\rho a^5}{4} = -\frac{ma^2}{4}$$

$$I_{zx} = -\iiint \rho zx\, \mathrm{d}x\, \mathrm{d}y\, \mathrm{d}z$$

$$= -\rho \int_0^a \mathrm{d}y \int_0^a z\, \mathrm{d}z \int_0^a x\, \mathrm{d}x = -\frac{\rho a^5}{4} = -\frac{ma^2}{4}$$

$$I_{zy} = -\iiint \rho zy\, \mathrm{d}x\, \mathrm{d}y\, \mathrm{d}z$$

$$= -\rho \int_0^a \mathrm{d}x \int_0^a z\, \mathrm{d}z \int_0^a y\, \mathrm{d}y = -\frac{\rho a^5}{4} = -\frac{ma^2}{4}$$

$$I_{zz} = \iiint \rho(x^2 + y^2)\, \mathrm{d}x\, \mathrm{d}y\, \mathrm{d}z$$

$$= \rho \int_0^a \mathrm{d}y \int_0^a \mathrm{d}z \int_0^a x^2\, \mathrm{d}x + \rho \int_0^a \mathrm{d}x \int_0^a \mathrm{d}z \int_0^a y^2\, \mathrm{d}y$$

$$= \frac{\rho a^5}{3} + \frac{\rho a^5}{3} = \frac{2\rho a^5}{3} = \frac{2ma^2}{3}$$

3.11.1 Principal axes

It is a remarkable fact, not immediately obvious here, that it is possible to orient a coordinate system fixed with the body in such a manner that all the products of inertia vanish. For bodies having symmetries, such as cylinders, rectangular prisms, it is easy to see, but it is true for any rigid body and for any point of body chosen as origin. The sets of coordinate axes for which the products of inertia vanish are known as the principal axes for the rigid body. We have $I_{xx} = I_1$, $I_{yy} = I_2$, $I_{zz} = I_3$.

Let us assume that a set of principal axes has been identified in the body, rotating with it and that we express the angular velocity as

$$\vec{\omega} = \omega_1 \hat{\imath} + \omega_2 \hat{\jmath} + \omega_3 \hat{k}$$

Here, $\hat{\imath}, \hat{\jmath}, \hat{k}$ are unit vectors along the principal axes and the components of $\vec{\omega}$ relate to these axes. The angular momentum vector is

$$\vec{J} = I_1\omega_1 \hat{\imath} + I_2\omega_2 \hat{\jmath} + I_3\omega_3 \hat{k} \tag{3.70}$$

As the coordinates are fixed with the body, the moments of inertia are constants, but the angular velocity components ω_1, ω_2, ω_3 may vary. Further, the unit vectors $\hat{\imath}$, $\hat{\jmath}$, $\hat{k}$ are changing with time as they rotate with the body. Taking time derivative of equation (3.70), we get

$$\frac{\mathrm{d}\vec{J}}{\mathrm{d}t} = I_1 \frac{\mathrm{d}\omega_1}{\mathrm{d}t} \hat{\imath} + I_2 \frac{\mathrm{d}\omega_2}{\mathrm{d}t} \hat{\jmath} + I_3 \frac{\mathrm{d}\omega_3}{\mathrm{d}t} \hat{k} + I_1\omega_1 \frac{\mathrm{d}\hat{\imath}}{\mathrm{d}t}$$
$$+ I_2\omega_2 \frac{\mathrm{d}\hat{\jmath}}{\mathrm{d}t} + I_3\omega_3 \frac{\mathrm{d}\hat{k}}{\mathrm{d}t} \tag{3.71}$$

Now, the unit vectors vary only because they rotate with angular velocity $\vec{\omega}$, and their rates of change are[7]

$$\frac{\mathrm{d}\hat{\imath}}{\mathrm{d}t} = \vec{\omega} \times \hat{\imath} \qquad \frac{\mathrm{d}\hat{\jmath}}{\mathrm{d}t} = \vec{\omega} \times \hat{\jmath} \qquad \frac{\mathrm{d}\hat{k}}{\mathrm{d}t} = \vec{\omega} \times \hat{k} \tag{3.72}$$

Using equations (3.72) in (3.71), we get

$$\frac{\mathrm{d}\vec{J}}{\mathrm{d}t} = I_1 \frac{\mathrm{d}\omega_1}{\mathrm{d}t} \hat{\imath} + I_2 \frac{\mathrm{d}\omega_2}{\mathrm{d}t} \hat{\jmath} + I_3 \frac{\mathrm{d}\omega_3}{\mathrm{d}t} \hat{k} + I_1\omega_1\, \vec{\omega} \times \hat{\imath}$$
$$+ I_2\omega_2\, \vec{\omega} \times \hat{\jmath} + I_3\omega_3\, \vec{\omega} \times \hat{k}$$

[7]The time-derivative operator can be expressed in case of rotational motion as

$$\frac{\mathrm{d}}{\mathrm{d}t} = \vec{\omega} \times$$

$$= I_1 \frac{d\omega_1}{dt}\,\hat{i} + I_2 \frac{d\omega_2}{dt}\,\hat{j} + I_3 \frac{d\omega_3}{dt}\,\hat{k} + \vec{\omega} \times (I_1\omega_1\,\hat{i} + I_2\omega_2\,\hat{j} + I_3\omega_3\,\hat{k})$$

$$= I_1 \frac{d\omega_1}{dt}\,\hat{i} + I_2 \frac{d\omega_2}{dt}\,\hat{j} + I_3 \frac{d\omega_3}{dt}\,\hat{k} + \vec{\omega} \times \vec{J}$$

If $\vec{\tau}$ be the external torque working on the body, we have

$$\vec{\tau} = \frac{d\vec{J}}{dt} = I_1 \frac{d\omega_1}{dt}\hat{i} + I_2 \frac{d\omega_2}{dt}\,\hat{j} + I_3 \frac{d\omega_3}{dt}\,\hat{k} + \vec{\omega} \times \vec{J} \tag{3.73}$$

We have

$$\vec{\omega} \times \vec{J} = \begin{vmatrix} \hat{i} & \hat{j} & \hat{k} \\ \omega_1 & \omega_2 & \omega_3 \\ I_1\omega_1 & I_2\omega_2 & I_3\omega_3 \end{vmatrix}$$

$$= \omega_2\omega_3(I_3 - I_2)\hat{i} + \omega_3\omega_1(I_1 - I_3)\hat{j} + \omega_1\omega_2(I_2 - I_1)\hat{k} \tag{3.74}$$

Expressing

$$\vec{\tau} = \tau_1\hat{i} + \tau_2\hat{j} + \tau_z\hat{k} \tag{3.75}$$

From equations (3.73), (3.74) and (3.75), we get

$$\tau_1 = I_1 \frac{d\omega_1}{dt} + \omega_2\omega_3(I_3 - I_2) \tag{3.76}$$

$$\tau_2 = I_2 \frac{d\omega_2}{dt} + \omega_3\omega_1(I_1 - I_3) \tag{3.77}$$

$$\tau_3 = I_3 \frac{d\omega_3}{dt} + \omega_1\omega_2(I_2 - I_1) \tag{3.78}$$

These are known as the Euler's equations.

3.11.2 Classification of rigid bodies

Symmetrical bodies may be classified into the following categories:

1. When $I_1 \neq I_2 \neq I_3$, the body is said to be asymmetric top.

2. When $I_1 \neq I_2 = I_3$ or $I_1 = I_2 \neq I_3$, the body is said to be symmetric top.

3. When $I_1 = I_2 = I_3$, the body is said to be spherical top.

Special cases

(a) When all the particles of a rigid body are restricted in a plane, their one coordinate can be taken as zero and we get

$$I_3 = I_1 + I_2$$

Here, I_3 is the moment of inertia about the axis perpendicular to the plane of the paper.

(b) When all the particles of a rigid body are collinear, two of their coordinates can be taken as zero and we have

$$I_1 = I_2 \qquad \text{and} \qquad I_3 = 0$$

Such a system is known as rotator and it has only two rotational degrees of freedom, as the rotation about the axis connecting all points of the rigid body is meaningless.

3.11.3 Free motion

For free motion, the net external torque acting on the body is zero, that is $\vec{\tau} = 0$ and Euler equations (3.76) – (3.78) reduce to

$$\frac{d\omega_1}{dt} - \frac{I_2 - I_3}{I_1}\omega_2\omega_3 = 0 \tag{3.79}$$

$$\frac{d\omega_2}{dt} - \frac{I_3 - I_1}{I_2}\omega_3\omega_1 = 0 \tag{3.80}$$

$$\frac{d\omega_3}{dt} - \frac{I_1 - I_2}{I_3}\omega_1\omega_2 = 0 \tag{3.81}$$

Free motion of symmetric top

Let us apply these Euler's equations for motion of a symmetric top. If x', y', z' are axes of orthogonal coordinates system attached to the top and z' is the axis of symmetry of the top, the moments of inertia about x' and y' are equal ($I_1 = I_2$). Thus, equation (3.81) gives

$$\frac{d\omega_3}{dt} = 0$$

It shows that ω_3 remains constant. Now, equations (3.79) and (3.80) can be expressed as

$$\frac{d\omega_1}{dt} = -\alpha\omega_2 \tag{3.82}$$

and

$$\frac{d\omega_2}{dt} = \alpha\omega_1 \tag{3.83}$$

where $\alpha = (I_3 - I_1)\omega_3/I_1$ is a constant. Multiplying equation (3.83) by $i = \sqrt{-1}$ and adding the product to equation (3.82), we get

$$\frac{d}{dt}(\omega_1 + i\omega_2) = i\alpha(\omega_1 + i\omega_2) \tag{3.84}$$

The solution of equation (3.84) is

$$(\omega_1 + i\omega_2) = A\,e^{i\alpha t} = A(\cos\alpha t + i\sin\alpha t) \tag{3.85}$$

where A is a constant, which may be made real by a suitable choice of the origin of time. Thus, we have

$$\omega_1 = A\cos\alpha t \qquad \text{and} \qquad \omega_2 = A\sin\alpha t \tag{3.86}$$

Obviously, $A = \sqrt{\omega_1^2 + \omega_2^2}$. Equation (3.86) shows that the components of angular velocity perpendicular to the axis of symmetry of the top rotates with an angular velocity α and have a constant amplitude A. Since ω_3 is constant, it can be concluded that the vector $\vec{\omega}$ rotates uniformly with the angular velocity α about the axis of symmetry of the top and its magnitude remains constant. Since the components of angular momentum $\vec{J}$ are

$$J_1 = I_1\omega_1 \qquad J_2 = I_2\omega_2 \qquad J_3 = I_3\omega_3$$

Thus, the angular momentum $\vec{J}$ also rotates uniformly with angular velocity α about the axis of symmetry of the top, remaining unchanged in magnitude.

Multiple-choice questions

1. Which of the following quantities is not conserved for inelastic collision but is conserved for elastic collision?

   (i) Mass (ii) Momentum

   (iii) Kinetic energy (iv) None of them

   Ans. (iii)

2. In an elastic collision between two equal masses, one of them is scattered through an angle θ when measured in laboratory coordinate system. When the scattering process is viewed in the centre-of-mass system, the scattering angle is ϕ. The relation between θ and ϕ is

   (i) $\theta = \phi/4$ (ii) $\theta = \phi/2$

   (iii) $\theta = \phi$ (iv) $\theta = 2\phi$

   Ans. (ii)

3. Position vectors of two equal masses are $\vec{r}_1 = 2\hat{i} + 3\hat{j} + 4\hat{k}$ and $\vec{r}_2 = \hat{i} - 2\hat{j} - 2\hat{k}$. The position vector of the centre of mass is

(i) $1.5\hat{i}+0.5\hat{j}+\hat{k}$ (ii) $0.5\hat{i}+2.5\hat{j}+3\hat{k}$

(iii) $-0.5\hat{i}-2.5\hat{j}-3\hat{k}$ (iv) $2\hat{i}-0.25\hat{j}-3\hat{k}$

Ans. (i)

4. A system has three bodies of masses 1, 2 and 2 units situated at the positions $\vec{r}_1 = 3\hat{i}+2\hat{j}+\hat{k}$, $\vec{r}_2 = \hat{i}+\hat{j}-\hat{k}$ and $\vec{r}_3 = -2\hat{i}+\hat{j}+2\hat{k}$, respectively. The position vector of the centre of mass is

(i) $\hat{i}+6\hat{j}+3\hat{k}$ (ii) $0.2\hat{i}+1.2\hat{j}+0.6\hat{k}$

(iii) $2\hat{i}+4\hat{j}+2\hat{k}$ (iv) $0.4\hat{i}+0.8\hat{j}+0.4\hat{k}$

Ans. (ii)

5. A system has three bodies of masses 2, 1 and 2 units moving with velocities $\vec{v}_1 = -\hat{i}+\hat{j}+2\hat{k}$, $\vec{v}_2 = 2\hat{i}+\hat{j}-2\hat{k}$ and $\vec{v}_3 = \hat{i}-2\hat{j}+\hat{k}$, respectively. The velocity of the centre of mass is

(i) $2\hat{i}-\hat{j}+4\hat{k}$ (ii) $2\hat{i}+\hat{k}$

(iii) $0.4\hat{i}-0.2\hat{j}+0.8\hat{k}$ (iv) $0.4\hat{i}+0.2\hat{k}$

Ans. (iii)

6. In xy-plane, three particles of masses 2, 3 and 1 units are placed at the points (3,4), (2, −1), (2,2), respectively. The moment of inertia of the system about x-axis in units of moment of inertia is

(i) 7 (ii) 14

(iii) 34 (iv) 39

Ans. (iv)

7. In xy-plane, three particles of masses 1, 2 and 3 units are placed at the points (4,5), (3, −2), (2,3), respectively. The moment of inertia of the system about z-axis in units of moment of inertia is

(i) 106 (ii) 60

(iii) 46 (iv) 14

Ans. (i)

8. For a body of mass m, rotating about an axis, the moment of inertia is I. If at any instant t, its linear velocity is $\vec{v}$ and angular velocity $\vec{\omega}$, the angular kinetic energy is

(i) $mv^2/2$ (ii) $m\,\vec{v}$

(iii) $I\omega^2/2$ (iv) $I\,\vec{\omega}$

Ans. (iii)

9. A particle of mass m is moving with instantaneous velocity $\vec{v}$ and angular velocity $\vec{\omega}$. If $\vec{r}$ be the position vector of the particle, which of the following relations is correct?

(i) $\vec{\omega} = \vec{r} \times \vec{v}$ (ii) $\vec{v} = \vec{\omega} \times \vec{r}$

(iii) $\vec{r} = \vec{v} \times \vec{\omega}$ (iv) $\vec{v} = \vec{r} \times \vec{\omega}$

Ans. (ii)

10. If I be the moment of inertia of a ring about an axis passing through its centre and perpendicular to its plane. The moment of inertia of the ring about an axis passing through its centre and lying in its plane is

(i) $I/2$ (ii) $I/3$

(iii) $I/4$ (iv) $2I$

Ans. (i)

11. If I be the moment of inertia of a disc about an axis passing through its centre and perpendicular to its plane. The moment of inertia of the disc about an axis passing through its centre and lying in its plane is

(i) $I/2$ (ii) $I/3$

(iii) $I/4$ (iv) $2I$

Ans. (i)

12. If I_1, I_2, I_3 are moments of inertia of a body about principal inertial axes. For asymmetrical top, we have

(i) $I_1 \neq I_2 \neq I_3$ (ii) $I_1 \neq I_2 = I_3$

(iii) $I_1 = I_2 = I_3$ (iv) $I_1 = I_2 \neq I_3$

Ans. (i)

13. If I_1, I_2, I_3 are moments of inertia of a body about principal inertial axes. For spherical top, we have

(i) $I_1 \neq I_2 \neq I_3$ (ii) $I_1 \neq I_2 = I_3$

(iii) $I_1 = I_2 = I_3$ (iv) $I_1 = I_2 \neq I_3$

Ans. (iii)

14. If I_1, I_2, I_3 are moments of inertia of a body about principal inertial axes. For symmetrical top, we have

(i) $I_1 \neq I_2 \neq I_3$ (ii) $I_1 \neq I_2 = I_3$

(iii) $I_1 = I_2 = I_3$ (iv) $I_1 \leq I_2 \leq I_3$

Ans. (ii)

Exercises

1. Describe the elastic and inelastic collisions of two bodies. Which physical quantities remain conserved and which do not remain conserved during both kinds of collisions?

2. Two bodies of masses m_1 and m_2 are moving with velocities $\vec{v}_1$ and $\vec{v}_2$, respectively, in the laboratory system of reference. After collision, they combine and move with velocity $\vec{v}$. Write down the equations for conservation of energy and momentum for elastic collision.

3. Describe the laboratory coordinate system and centre-of-mass system. A body 1 of mass m_1 moving along x-axis with velocity $v_1\hat{i}$ collides with another body 2 of mass m_2 which is stationary in the laboratory coordinate system. Show that when we use centre-of-mass system, the momenta of the bodies 1 and 2 are opposite to each other.

4. Two bodies 1 and 2 of masses $m_1 = 0.5$ kg and $m_2 = 0.8$ kg, moving with velocities $\vec{v}_1 = 2\hat{i} - 3\hat{j} + 4\hat{k}$ and $\vec{v}_2 = 3\hat{i} + 2\hat{j} - 4\hat{k}$, respectively, collide with each other. After collision, body 1 moves with velocity $\vec{v}_1' = 3\hat{i} - 2\hat{j} + 4\hat{k}$. Calculate the velocity of body 2.

5. Two bodies of masses m_1 and m_2 collide with each other. Show that when the two masses are equal, the scattering angle in the laboratory system of reference is half of that in the centre-of-mass system.

6. Two bodies of masses m_1 and m_2 collide with each other. Show that when $m_1 << m_2$, the scattering angle in the laboratory system of reference is equal to that in the centre-of-mass system.

7. A particle of mass m moving with velocity v collides with another particle of equal mass m which is stationary. Show that the scattering angle of moving particle can neither be 0° nor 90°.

8. Show that total torque due to internal forces is zero.

9. Show that for elastic collision of two bodies in the centre-of-mass system, the velocities of particles remain unchanged and only the direction of motion changes.

10. Derive expression for centre of mass.

11. Suppose four masses 1.5 kg, 2.5 kg, 3 kg and 4 kg are situated at the positions given by $\vec{r}_1 = -2\hat{i} + \hat{j} + 4\hat{k}$, $\vec{r}_2 = 3\hat{i} + 2\hat{j} + \hat{k}$, $\vec{r}_3 = 3\hat{i} + \hat{j} - 2\hat{k}$ and $\vec{r}_4 = 3\hat{i} + 5\hat{j} + \hat{k}$, respectively, in a laboratory system of reference. Calculate the position vector of the centre of mass of the system.

12. Suppose three masses 2 kg, 3 kg and 4 kg are moving with velocities $\vec{v}_1 = 3\hat{i} - 2\hat{j} + \hat{k}$, $\vec{v}_2 = -2\hat{i} + 3\hat{j} + 2\hat{k}$ and $\vec{v}_3 = 4\hat{i} - 2\hat{j} + \hat{k}$, respectively, in a laboratory system of reference. Calculate the velocity of the centre of mass of the system.

13. Two bodies of masses m_1 and m_2 are moving with velocities $\vec{v}_1$ and $\vec{v}_2$, respectively, in the laboratory system of reference. Show that in the centre-of-mass system, the momenta of the bodies are opposite to each other.

14. Two bodies 1 and 2 of masses m_1 and m_2 are moving with velocities $\vec{v}_1$ and $\vec{v}_2$, respectively, in a laboratory frame of reference. Show that in a centre-of-mass system, the momenta of the bodies are equal and opposite to each other.

15. Define angular momentum and torque. Show that the rate of change of angular momentum of a body is equal to the external torque applied on the body.

16. Define angular momentum of a system of bodies. In a system, three bodies of masses 0.5 kg, 2 kg and 1 kg are moving with velocities $\vec{v}_1 = \hat{\imath} + 2\hat{\jmath} + \hat{k}$, $\vec{v}_2 = -\hat{\imath} + \hat{\jmath} + 2\hat{k}$ and $\vec{v}_3 = 2\hat{\imath} + 2\hat{\jmath} - \hat{k}$, respectively. Calculate total angular momentum of the system when the positions of the masses are $(1, -1, 2)$, $(1, 2, 2)$ and $(1, 3, -2)$, respectively.

17. Define angular momentum of a system of bodies about its centre of mass. In a system, two bodies of masses 1 kg and 2 kg are moving with velocities $\vec{v}_1 = 2\hat{\imath} - \hat{\jmath} + 3\hat{k}$ and $\vec{v}_2 = -\hat{\imath} + 2\hat{\jmath} - 2\hat{k}$, respectively. Calculate total angular momentum of the system about its centre of mass when the positions of the masses are, respectively, $(1, -3, 2)$ and $(1, -1, 3)$.

18. Derive relation between angular momentum $\vec{J}$ and torque $\vec{\tau}$.

19. Show that total torque due to internal forces is zero.

20. Show that for a central force, the torque is zero and angular momentum is conserved.

21. Suppose a planet of mass m revolves around the sun in a circular orbit of radius r with velocity v. Derive an expression for angular momentum for revolution of planet around the sun.

22. Suppose a planet of mass m rotates about its own axis with time period T. If R is the radius of spherical planet, derive an expression for angular momentum for rotation of the planet.

23. Consider a mass revolving around an axis. What happens when the mass comes closer to the axis of rotation without applying any external torque?

24. For a rotational motion along a circular path of a body, show that angular momentum J can be expressed as $J = I\omega$, where I is the moment of inertia of body about the axis of rotation and ω the angular velocity.

25. Describe the process for formation of a galaxy. How the mass of our galaxy, Milky Way, can be estimated?

26. Show that the mass of our galaxy, Milky Way, can be estimated if we know distance of our sun from the centre of Milky Way and time period of its revolution around the centre of the Milky Way.

27. Describe angular motions of a planet in our solar system. How the mass of the sun can be estimated?

28. Show that the mass of sun can be estimated if we know distance of a planet from the sun and time period of the revolution of the planet around the sun.

29. Obtain expressions for angular momentum due to revolution and due to rotation of a planet in our solar system.

30. Obtain expressions for angular momentum and kinetic energy of a particle of mass m rotating with angular velocity ω about an axis.

31. Obtain expressions for angular momentum and kinetic energy of a body of mass M, consisting of a large number of particles and rotating with angular velocity ω about an axis.

32. Define moment of inertia of a body consisting a number of particles.

33. State and prove the theorem of parallel axis for moment of inertia.

34. State and prove the theorem of perpendicular axis for moment of inertia.

35. In a plane lamina of mass 8 kg, there are three points A, B, C forming a triangle, as shown below. The point C is the centre of mass of the lamina. Three parallel axes, perpendicular to the lamina are passing through A, B and C. If the moment of inertia of the body about the axis passing through the point A is 600 kg-m^2, calculate the moment of inertia of the body about the axis passing through the point B.

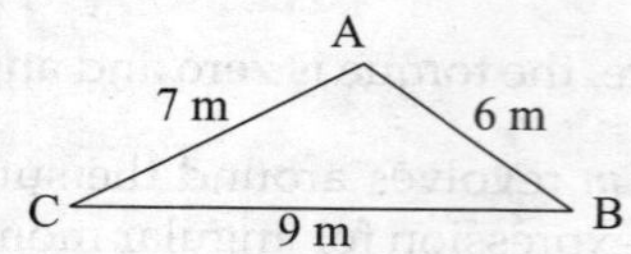

36. Let us consider a uniform rectangular lamina of length a, width b and mass m. Calculate moment of inertia of the lamina about x-axis (parallel to the length), passing through the centre of mass and lying in the plane of the lamina.

37. Two equal masses connected by a light mass rod rotate about a fixed axis passing through the centre of mass and oriented at an angle θ from the rod with angular velocity ω. Show that the angular momentum of the system is not constant. Also show that the time derivative of the angular momentum is in the direction perpendicular to axis of rotation.

38. Describe the moments and products of inertia of a body.

39. Write notes on the following:

(i) Conservation laws
(ii) Elastic and inelastic collisions
(iii) Centre of mass of a system of bodies
(iv) Angular momentum
(v) Torque
(vi) Momentum and its conservation
(vii) Angular momentum of a system of bodies
(viii) Angular momentum of a system of bodies about its centre of mass
(ix) Determination of mass of our galaxy, Milky Way

(x) Angular motions of a planet in our solar system

(xi) Angular momentum about the centre of mass

(xii) Angular momentum for revolution of a planet around the sun

(xiii) Angular momentum for rotation of a planet about its own axis

(xiv) Formation of a galaxy

(xv) Estimation of mass of the sun

(xvi) Theorem of parallel axis for moment of inertia

(xvii) Theorem of perpendicular axis for moment of inertia

(xviii) Principal axes of a rigid body

(xix) Euler angles

CHAPTER 4

INVERSE-SQUARE-LAW FORCES

A force which varies as an inverse of the square of the separation between two bodies is known as the inverse-square-law force. Examples of such force are the gravitational force between two masses, electrostatic force between two charges and magnetic force between two magnetic poles. It is the gravitational force which controls motion of heavenly bodies. In this chapter, we shall discuss about the inverse-square-law forces.

4.1 Central force

A force which is always directed towards a point (generally called the centre) and is proportional to the distance of the body from the point is known as the central force. It may be expressed as

$$\vec{F} = f(r)\hat{r}$$

where $f(r)$ is function of distance r and $\hat{r}$ a unit vector along the line joining the centre of force and the body on which the force is acting. Examples of such force are (i) force on electron revolving around a nucleus, (ii) force on a planet revolving around the sun, (iii) force on a satellite revolving around the respective planet, (iv) force between two nucleons in a nucleus, etc. We can say that the force between two charges and between two masses are examples of the central force. The torque τ for the central force is

$$\vec{\tau} = \vec{r} \times \vec{F} = \vec{r} \times f(r)\hat{r} = 0$$

It shows that the torque for a central force is zero. For the zero value of torque, equation

$$\vec{\tau} = \frac{\mathrm{d}\vec{J}}{\mathrm{d}t}$$

says that the angular momentum is conserved. Hence, for a central force the angular momentum is conserved.

The gravitational force $\vec{F}$ between two masses m_1 and m_2 separated by a distance r is

$$\vec{F} = -\frac{Gm_1m_2}{r^2}\hat{r}$$

As the gravitational force is always attractive, a negative sign is used. Here, $\hat{r}$ is unit vector along the direction of the gravitational force. This is an example of central force, we have:

$$f(r) = -\frac{Gm_1m_2}{r^2}$$

Another example of central force is the force between an electron and a nucleus. Suppose q_1 is charge on electron and q_2 on the nucleus, separated by a distance r. The Coulomb force between them is

$$\vec{F} = \frac{1}{4\pi\epsilon_0}\frac{q_1q_2}{r^2}\hat{r} \qquad \text{so that} \qquad f(r) = \frac{1}{4\pi\epsilon_0}\frac{q_1q_2}{r^2}$$

In a nucleus, proton as well as neutron is known as a nucleon. The force between two protons, between two neutrons, between one proton and one neutron is known as a nuclear force. The nuclear force also depends only on the separation between two nucleons. Thus, the nuclear force also is a central force.

Exercise 4.1: A planet of mass m is revolving around the sun having mass M in a circular path of radius r. Show that the angular momentum of revolution is constant.

Solution: The gravitational force acting on the planet due to the sun is

$$\vec{F} = -\frac{GmM}{r^2}\hat{r}$$

Thus, the torque of the force is

$$\vec{\tau} = \vec{r} \times \vec{F} = -\vec{r} \times \frac{GmM}{r^2}\hat{r} = 0$$

as $\vec{r} \times \hat{r} = 0$. The relation between the angular momentum $\vec{J}$ and external torque is

$$\vec{\tau} = \frac{d\vec{J}}{dt}$$

For zero value of torque, we have

$$\frac{d\vec{J}}{dt} = 0 \qquad \text{or} \qquad \vec{J} = \text{constant}$$

It shows that the angular momentum for the revolution of a planet around the sun is conserved.

Exercise 4.2: An electron of charge q_1 is revolving around nucleus having charge q_2 in a circular path of radius r. Show that the angular momentum of revolution of electron is constant.

Solution: The force acting on an electron due to the nucleus is

$$\vec{F} = \frac{1}{4\pi\epsilon_0}\frac{q_1q_2}{r^2}\hat{r}$$

Thus, the torque of the force is

$$\vec{\tau} = \vec{r} \times \vec{F} = \vec{r} \times \frac{1}{4\pi\epsilon_0} \frac{q_1 q_2}{r^2} \hat{r} = 0$$

as $\vec{r} \times \hat{r} = 0$. The relation between the angular momentum $\vec{J}$ and external torque is

$$\vec{\tau} = \frac{d\vec{J}}{dt}$$

For zero value of torque, we have

$$\frac{d\vec{J}}{dt} = 0 \qquad \text{or} \qquad \vec{J} = \text{constant}$$

It shows that the angular momentum for the revolution of an electron around the nucleus is conserved.

4.2 Inverse-square-law force

A force which is proportional to r^{-2} is known as the inverse-square-law force. It is one form of central force. Thus, the force is

$$F = \frac{C}{r^2}$$

The central force is directed along its line of action. If one particle is at the origin and the second particle is at a position $\vec{r}$, the force on the second particle is

$$\vec{F} = \frac{C}{r^2} \hat{r}$$

where for gravitational force between two masses m_1 and m_2, we have

$$C = -Gm_1 m_2$$

and for Coulomb force between two charges q_1 and q_2, we have

$$C = \frac{q_1 q_2}{4\pi\epsilon_0}$$

In the spherical polar coordinates when the distribution of physical quantities depends only on the radial distance r, the relation between the force $\vec{F}$ and potential V is[1]

$$\vec{F} = -\frac{dV}{dr} \hat{r} \tag{4.1}$$

Notice that potential is a scalar quantity. Thus, for the inverse-square-law force, we have

$$\frac{C}{r^2} = -\frac{dV}{dr} \qquad \text{or} \qquad dV = -\frac{C}{r^2} dr$$

[1]The force $\vec{F}$ and potential V have a relation $\vec{F} = -\nabla V$.

On integration of this equation, we have

$$V = \frac{C}{r} + c_1$$

where c_1 is constant of integration. Considering that $V = 0$ when $r = \infty$, we get $c_1 = 0$ and therefore

$$V = \frac{C}{r}$$

Hence, the gravitational potential and electrostatic potential are

$$V = -\frac{Gm_1m_2}{r} \quad \text{and} \quad V = \frac{1}{4\pi\epsilon_0}\frac{q_1q_2}{r}$$

respectively. As potential is a scalar quantity and addition of two scalars is easier as compared to the addition of two vectors, we shall calculate first potential and then obtain the corresponding force with the help of relation (4.1).

4.3 Potential between a point mass and a spherical shell

Let us consider a spherical shell of mass M and radius R. Thus, the mass per unit surface area of the shell is

$$\sigma = \frac{M}{4\pi R^2}$$

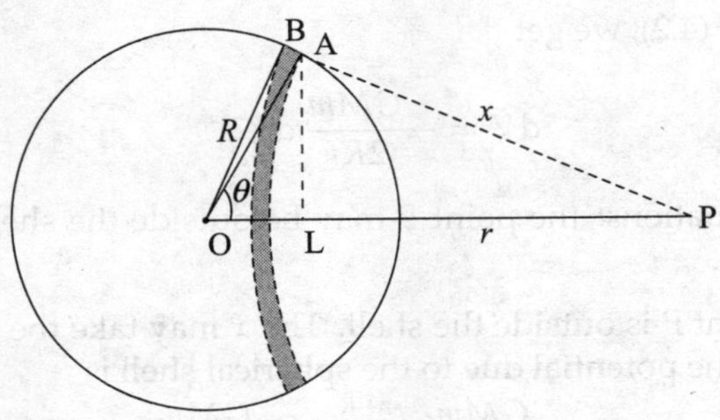

Figure 4.1: Spherical shell and a point P outside the shell.

Suppose a point mass m is placed at a point P which is at a distance r from the centre O of the shell (Figure 4.1). This point P is outside the shell ($r > R$). It may be inside the shell ($r < R$). For calculation of potential between the mass m and spherical shell, let us consider a ring on the shell having angular width $\mathrm{d}\theta$ so that AB $= R\,\mathrm{d}\theta$. The radius of the ring is AL $= R\sin\theta$ and therefore the circumference of the ring is $2\pi R\sin\theta$. The area of the ring is

$$(2\pi R\sin\theta)(R\,\mathrm{d}\theta) = 2\pi R^2 \sin\theta\,\mathrm{d}\theta$$

and mass of the ring is

$$m_r = 2\pi R^2 \sin\theta\,\mathrm{d}\theta\,\sigma = \frac{M\sin\theta\,\mathrm{d}\theta}{2}$$

Here, we used the value of the surface density σ. Each point on the ring is at a distance x from the point P.

The ring may be considered consisting of a large number of small elements so that $m_r = \sum m_i$, where m_i is the mass of the ith element. The potential between two masses m_i and m is

$$dV_i = -G\,\frac{m_i m}{x}$$

Thus, the potential between the ring and the mass m can be obtained by adding the potentials due to all elements; the potential is

$$dV = \sum dV_i = -G\,\frac{m \sum m_i}{x}$$

$$= -G\frac{m m_r}{x} = -G\frac{Mm \sin\theta \, d\theta}{2x} \tag{4.2}$$

Here, we used the expression for m_r. By law of cosine for the triangle OAP in Figure 4.1, we have

$$\cos\theta = \frac{r^2 + R^2 - x^2}{2rR} \qquad \text{or} \qquad x^2 = r^2 + R^2 - 2rR\cos\theta$$

On differentiation, we get

$$x\,dx = rR\sin\theta\,d\theta \qquad \text{or} \qquad \sin\theta\,d\theta = \frac{x\,dx}{rR} \tag{4.3}$$

Using equation (4.3) in (4.2), we get

$$dV = -\frac{GMm}{2Rr}\,dx$$

Now, we have two situations: the point P may be outside the shell ($r > R$) and may be inside the shell ($r < R$).

(i) Suppose the point P is outside the shell. The x may take the values from $(r - R)$ to $(r + R)$ and therefore the potential due to the spherical shell is

$$V = -\frac{GMm}{2Rr}\int_{r-R}^{r+R} dx = -\frac{GMm}{r}$$

It shows that the potential at a point outside the shell varies with the distance of the point from the centre of the shell. The force is

$$F = -\frac{dV}{dr} = -\frac{GMm}{r^2}$$

When the point P is on the surface of the shell ($r = R$). We have

$$V = -\frac{GMm}{R}$$

and the force is $F = 0$.

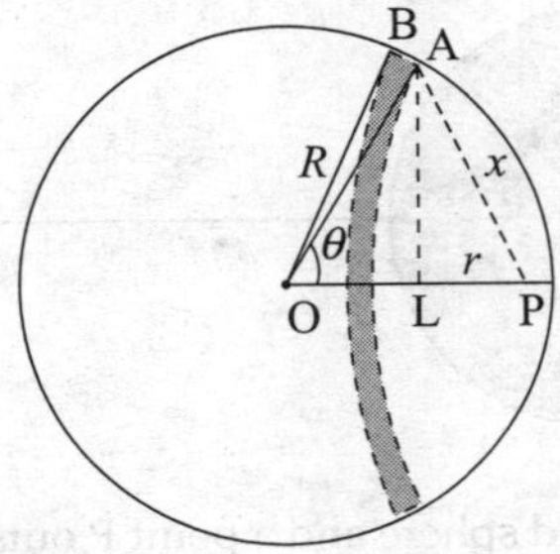

Figure 4.2: Spherical shell and a point P inside the shell.

(ii) Suppose the point P is inside the shell (Figure 4.2). The x may take the values from $(R-r)$ to $(r+R)$ and therefore the potential due to the spherical shell is

$$V = -\frac{GMm}{2Rr}\int_{R-r}^{r+R} \mathrm{d}x = -\frac{GMm}{R}$$

It shows that the potential at a point inside the shell does not depend on the position of the point. It is the same throughout inside the shell.

The above two cases can be jointly expressed as

$$V = \begin{cases} -GMm/r & r > R \\ -GMm/R & r < R \end{cases}$$

On differentiation of V, the force is

$$F = -\frac{\mathrm{d}V}{\mathrm{d}r} = \begin{cases} -GMm/r^2 & r > R \\ 0 & r < R \end{cases}$$

4.4 Potential between a point mass and a solid sphere

Let us consider a solid sphere of mass M and radius R. The mass per unit volume of the sphere is

$$\rho = \frac{M}{4\pi R^3/3}$$

Suppose a point mass m is placed at a point P which is at a distance r from the centre of the sphere. This point P is outside the sphere ($r > R$). It may be inside the sphere ($r < R$) as well.

(i) Suppose the point P is outside the solid sphere (Figure 4.3). In order to calculate the potential at P, let us divide the solid sphere into a large number of concentric spherical shells with a common centre at O, so that $\sum m_i = M$, where m_i denotes the mass of the ith

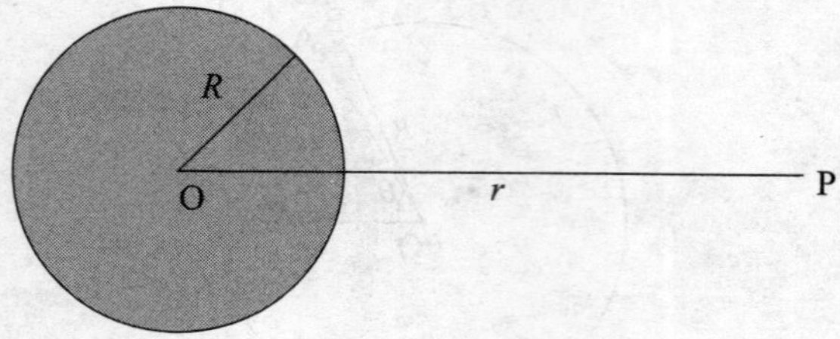

Figure 4.3: Solid sphere and a point P outside the sphere.

shell. Using the knowledge about the spherical shell, the potential at P due to the ith shell is

$$V_i = -\frac{Gm_i m}{r}$$

The potential V due to the solid sphere can be obtained by adding the potentials due to all the shells. That is

$$V = \sum V_i = -\frac{Gm\sum m_i}{r} = -\frac{GmM}{r}$$

The force at the point P due to the solid sphere is

$$F = -\frac{\mathrm{d}V}{\mathrm{d}r} = -\frac{GMm}{r^2}$$

When the point is on the surface of the sphere, the potential is

$$V = -\frac{GmM}{R}$$

and the force $F = 0$.

(ii) Suppose the point P is inside the solid sphere at a distance r from the centre of the sphere (Figure 4.4a). In order to calculate the potential at P, let us construct a sphere of radius r (Figure 4.4b). It divides the solid sphere into two parts: (a) the solid sphere of radius r and (b) a spherical shell of inner radius r and outer radius R.

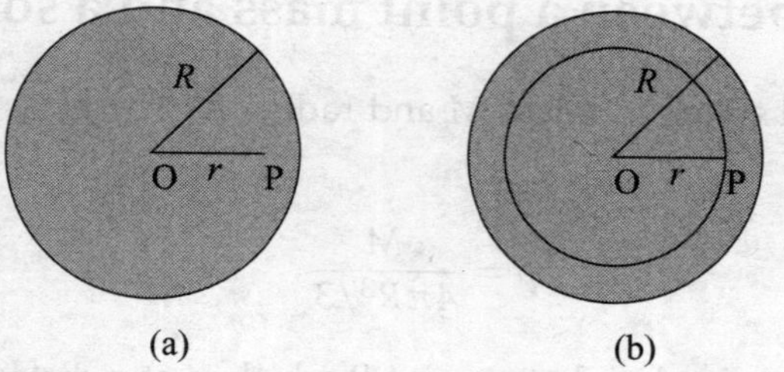

Figure 4.4: Solid sphere and a point P inside the sphere.

The mass of the solid sphere of radius r is

$$M' = \frac{4}{3}\pi r^3 \rho = \frac{Mr^3}{R^3}$$

Here, we used the value of ρ. Thus, the potential at P due to the mass M' is

$$V_1 = -\frac{GmM'}{r} = -\frac{GmMr^2}{R^3}$$

For computation of the potential due to the spherical shell of inner radius r and outer radius R, let us consider a spherical shell of radius x and thickness $\mathrm{d}x$ (Figure 4.5). The volume of the shell is

$$\text{volume} = \frac{4}{3}\pi(x+\mathrm{d}x)^3 - \frac{4}{3}\pi x^3 = 4\pi x^2\,\mathrm{d}x$$

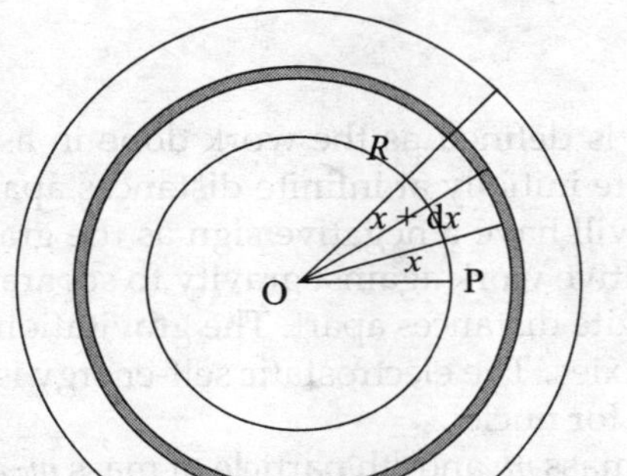

Figure 4.5: Spherical shell of radius x and thickness $x + \mathrm{d}x$.

The mass of the shell is

$$m_2 = 4\pi x^2 \rho\,\mathrm{d}x$$

The potential due the shell of mass m_2 at P, which is situated inside the shell, is

$$\mathrm{d}V_2 = -\frac{Gmm_2}{x} = -G4\pi xm\rho\,\mathrm{d}x \tag{4.4}$$

Here, we used the value of m_2. Thus, the potential due to the shell of inner radius r and outer radius R is obtained by integrating equation (4.4) as

$$V_2 = -G4\pi m\rho \int_r^R x\,\mathrm{d}x = -G2\pi m\rho(R^2 - r^2)$$

Using the value of ρ, we get

$$V_2 = -\frac{3GMm(R^2 - r^2)}{2R^3}$$

thus, the total potential due to the solid sphere is

$$V = V_1 + V_2 = -\frac{GmMr^2}{R^3} - \frac{3GMm(R^2 - r^2)}{2R^3}$$

$$= -\frac{GmM}{2R^3}(3R^2 - r^2)$$

and the force is

$$F = -\frac{\mathrm{d}V}{\mathrm{d}r} = -\frac{GmMr}{R^3}$$

When the point P coincides with the centre O, we have $r = 0$ and the potential is

$$V = -\frac{3GmM}{2R}$$

and the force is

$$F = -\frac{\mathrm{d}V}{\mathrm{d}r} = 0$$

4.5 Self-energy

The self-energy of a body is defined as the work done in assembling the body from infinitesimal elements that are initially at infinite distances apart. Let us consider the gravitational self-energy; this will have a negative sign as the gravitational force is attractive. That is, we have to do positive work against gravity to separate the particles of a body, for taking each particle to infinite distances apart. The gravitational self-energy is required for formation of stars and galaxies. The electrostatic self-energy is calculated for crystals, both insulators and metals, and for nuclei.

Suppose ith particle of mass m_i and jth particle of mass m_j are separated by the distance r_{ij}. The potential energy between them is

$$V_{ij} = -\frac{Gm_i m_j}{r_{ij}}$$

The potential energy of n discrete masses due to their mutual gravitational attraction is equal to the sum of the potential energy of all pairs of masses of particles:

$$V = -G\sum_{i=1}^{n}\sum_{j=1}^{n}{}' \frac{m_i m_j}{2r_{ij}} \tag{4.5}$$

Here, the prime denotes that j is not equal to i. A factor of 2 is introduced in the denominator, as each pair of particles is counted two times.

4.5.1 Gravitational energy of a galaxy

Let us estimate the gravitational energy of a galaxy. In the estimation, we shall not account for the gravitational energy of an individual star. We need to estimate the value of the expression in equation (4.5). Let us assume that the galaxy is made up of n individual stars. As an approximation, let us consider that each star has a mass M and separation between each pair of stars is of the order of R. Then from equation (4.5), we have

$$V = -G\sum_{i=1}^{n}\sum_{j=1}^{n}{}' \frac{M^2}{2R} = -\frac{G(n-1)nM^2}{2R}$$

Since the number n of stars is very large, we can take $(n-1) \approx n$ and therefore, we have

$$V = -\frac{Gn^2M^2}{2R}$$

Exercise 4.3: A galaxy has about 2×10^{11} stars, each of mass 2×10^{30} kg and radius 1.5×10^{21} m. Calculate gravitational self-energy of the galaxy. Given $G = 6.67 \times 10^{-11}$ N-m^2/kg^2.

Solution: The gravitational self-energy of the galaxy is

$$V = -\frac{Gn^2M^2}{2R} = -\frac{6.67 \times 10^{-11} \times (2 \times 10^{11})^2(2 \times 10^{30})^2}{2 \times 1.5 \times 10^{21}}$$

$$= -3.6 \times 10^{51}\text{ J}$$

4.5.2 Gravitational energy of a uniform sphere

We can easily calculate self-energy of a uniform sphere of mass M and radius R. The mass density of the sphere is

$$\rho = \frac{M}{4\pi R^3/3}$$

Let us consider that during the formation of a sphere of mass density ρ, at some instant the radius is x, so that $x < R$. The mass of the sphere is

$$M' = \frac{4}{3}\pi x^3 \rho$$

To this sphere, let us add a layer (spherical shell) of thickens $\mathrm{d}x$. The volume of the shell of inner radius x and outer radius $x + \mathrm{d}x$ is

$$V'' = \frac{4}{3}\pi(x+\mathrm{d}x)^3 - \frac{4}{3}\pi x^3 = 4\pi x^2\,\mathrm{d}x$$

The mass of the shell is

$$M'' = 4\pi\rho x^2\,\mathrm{d}x$$

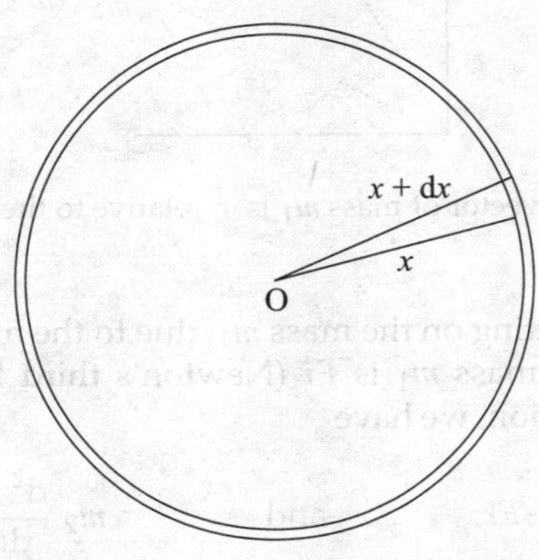

The potential between the sphere of radius x and shell of thickness $\mathrm{d}x$ is

$$\mathrm{d}V = -G\,\frac{(4\pi x^3\rho/3)(4\pi\rho x^2\,\mathrm{d}x)}{x} = -\frac{1}{3}\,G(4\pi\rho)^2 x^4\,\mathrm{d}x \tag{4.6}$$

The sphere is formed from the radius zero to R. Thus, the gravitational self-energy of a sphere of radius R can be obtained by integrating equation (4.6) for radius from zero to R. We have

$$V = -\frac{1}{3}\,G(4\pi\rho)^2 \int_0^R x^4\,\mathrm{d}x = -\frac{1}{3}\,G(4\pi\rho)^2\,\frac{R^5}{5}$$

$$= -\frac{3}{5}\,G\Big(\frac{4\pi}{3}\,R^3\rho\Big)^2\frac{1}{R} = -\frac{3}{5}\frac{GM^2}{R}$$

where $M = 4\pi R^3/3$ is the mass of sphere of radius R.

Exercise 4.4: Our sun has mass of 2×10^{30} kg and radius of 7×10^8 m. Calculate its gravitational self-energy. Given $G = 6.67\times10^{-11}$ N-m^2/kg^2.

Solution: The gravitational self-energy of the sun is

$$V = -\frac{3}{5}\frac{GM^2}{R} = -\frac{3\times6.67\times10^{-11}(2\times10^{30})^2}{5\times7\times10^8} = -2.3\times10^{41}\ \mathrm{J}$$

4.6 Two-body problem

We have solved the problem of one mass moving in the field of an infinitely large (or stationary) mass. Let us see what happens when two masses are comparable to each other. Suppose two masses m_1 and m_2 are situated at the points A and B whose position vectors are $\vec{r}_1$ and $\vec{r}_2$, respectively, as shown in Figure 4.6.

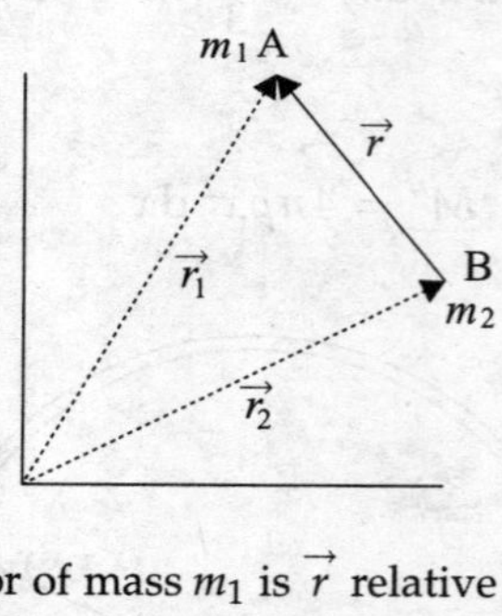

Figure 4.6: Position vector of mass m_1 is $\vec{r}$ relative to the position of mass m_2.

Suppose $(-F\hat{r})$ is the force acting on the mass m_1, due to the mass m_2, then the force acting on the mass m_2 due to the mass m_1 is $F\hat{r}$ (Newton's third law of motion). Then from Newton's second law of motion, we have

$$m_1\,\frac{\mathrm{d}^2\vec{r}_1}{\mathrm{d}t^2} = -F\hat{r} \qquad \text{and} \qquad m_2\,\frac{\mathrm{d}^2\vec{r}_2}{\mathrm{d}t^2} = F\hat{r}$$

or

$$\frac{\mathrm{d}^2\vec{r}_1}{\mathrm{d}t^2} = -\frac{F\hat{r}}{m_1} \qquad \text{and} \qquad \frac{\mathrm{d}^2\vec{r}_2}{\mathrm{d}t^2} = \frac{F\hat{r}}{m_2}$$

Subtracting the second equation from the first, we get

$$\frac{d^2(\vec{r}_1 - \vec{r}_2)}{dt^2} = -\left(\frac{1}{m_1} + \frac{1}{m_2}\right) F\hat{r} \tag{4.7}$$

We see that $\vec{r}_1 - \vec{r}_2 = \vec{r}$ is the position vector of m_1 relative to m_2 and the unit vector $\hat{r}$ is along $\vec{r}_1 - \vec{r}_2$. Now, we introduce the reduced mass μ of the system of masses m_1 and m_2 as

$$\frac{1}{\mu} = \frac{1}{m_1} + \frac{1}{m_2} \qquad \text{or} \qquad \mu = \frac{m_1 m_2}{m_1 + m_2}$$

Thus, equation (4.7) reduces to

$$\frac{d^2\vec{r}}{dt^2} = -\frac{F\hat{r}}{\mu} \qquad \text{or} \qquad \mu\,\frac{d^2\vec{r}}{dt^2} = -F\hat{r}$$

In case of a gravitational force between two masses m_1 and m_2, we have

$$F = G\,\frac{m_1 m_2}{r^2} \qquad \text{and} \qquad \mu\,\frac{d^2\vec{r}}{dt^2} = -G\,\frac{m_1 m_2}{r^2}\,\hat{r} \tag{4.8}$$

Equation (4.8) is used in the following way. Remember that $\vec{r}$ is a vector from m_2 to m_1. With the help of equation (4.8), we may discuss the motion of m_1 relative to m_2, exactly in a manner as if m_2 is fixed at the origin of an inertial frame, except that we must use μ instead of m_1 as the mass. Note that the force in equation (4.8) is not $-G\mu m_2/r^2$. Thus, to discuss the orbit of two-body problem, we only need to solve this one-body problem, expressed by equation (4.8). Equation (4.8) can be expressed as

$$\mu\,\frac{d^2\vec{r}}{dt^2} = -G\,\frac{\mu M}{r^2}\,\hat{r}$$

where $M = m_1 + m_2$ is total mass of the system of two masses. It shows that motion of two bodies of masses m_1 and m_2 is equivalent to the motion of one body of mass μ moving about other stationary body of mass M at a distance r from M.

The reduction of two-body problem to a one-body problem can be obtained in the same way for any central force; the reduced mass will always appear.

Reduced mass

For two masses m_1 and m_2, the reduced mass μ is expressed as

$$\frac{1}{\mu} = \frac{1}{m_1} + \frac{1}{m_2} \qquad \text{or} \qquad \mu = \frac{m_1 m_2}{m_1 + m_2}$$

Case 1: When one of the masses is much smaller than the other, say, $m_1 << m_2$, we have

$$\mu \approx \frac{m_1 m_2}{m_2} = m_1$$

It shows that when one of the masses is much smaller than the other, the reduced mass is equal to the smaller mass. For example, the reduced mass for one electron and one proton is equal to the mass of the electron.

Case 2: When the two masses are equal, $m_1 = m_2 = m$, we have

$$\mu = \frac{mm}{m+m} = \frac{m}{2}$$

It shows that when the two masses are equal, the reduced mass is equal to half of the mass of one particle. For example, reduced mass of H_2 molecule is equal to the half of the mass of hydrogen atom.

Exercise 4.5: Show that the reduced mass μ of two masses m_1 and m_2 is smaller than both m_1 and m_2.

Solution: The reduced mass μ of masses m_1 and m_2 is

$$\frac{1}{\mu} = \frac{1}{m_1} + \frac{1}{m_2} \qquad \text{or} \qquad \mu = \frac{m_1 m_2}{m_1 + m_2}$$

Suppose $m_1 < m_2$. We have

$$\mu = m_1 \frac{m_2}{m_1 + m_2}$$

Since, $m_2/(m_1 + m_2)$ is smaller than 1, we have $\mu < m_1$. Thus, μ is smaller than both m_1 and m_2.

Exercise 4.6: Masses of H and Cl atoms are 1 and 35.5 amu. Calculate the reduced mass of HCl molecule.

Solution: The reduced mass μ in HCl molecule is

$$\frac{1}{\mu} = \frac{1}{1} + \frac{1}{35.5} = \frac{35.5+1}{35.5} \qquad \text{or} \qquad \mu = \frac{35.5}{36.5} = 0.97 \text{ amu}$$

Exercise 4.7: Masses of C and O atoms are 12 and 16 amu. Calculate the reduced mass of CO molecule.

Solution: The reduced mass μ in CO molecule is

$$\frac{1}{\mu} = \frac{1}{12} + \frac{1}{16} = \frac{4+3}{48} \qquad \text{or} \qquad \mu = \frac{48}{7} = 6.86 \text{ amu}$$

4.6.1 Angular momentum and energy

Let us see how the angular momentum $\vec{J}$ and energy E of two-body system are expressed. Consider two bodies of masses m_1 and m_2 whose position vectors are $\vec{r}_1$ and $\vec{r}_2$, respectively, relative to the centre of mass, as shown in Figure 4.7. Then, we have

$$m_1 \vec{r}_1 = -m_2 \vec{r}_2$$

Here, the negative sign appears as the vectors $\vec{r}_1$ and $\vec{r}_2$ are directed in the opposite directions.

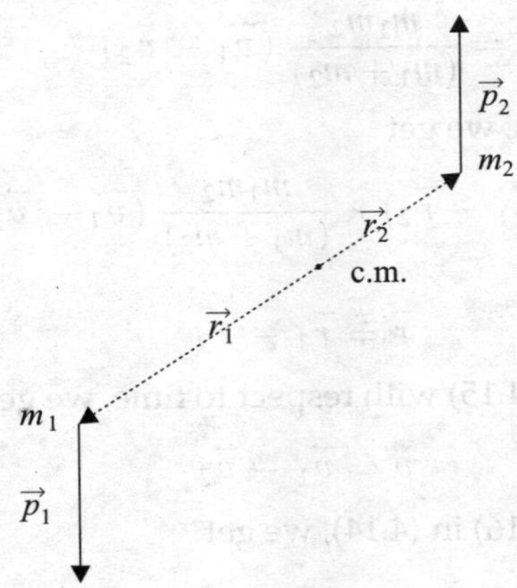

Figure 4.7: Relative to the centre of masses, the position vectors of m_1 and m_2 are $\vec{r}_1$ and $\vec{r}_2$, respectively.

On differentiating this equation with respect to time, we have

$$m_1 \frac{d\vec{r}_1}{dt} = -m_2 \frac{d\vec{r}_2}{dt} \qquad \text{or} \qquad m_1\vec{v}_1 = -m_2\vec{v}_2 \tag{4.9}$$

where $\vec{v}_1$ and $\vec{v}_2$ are, respectively, velocities of the bodies.

Angular momentum

Total angular momentum of the system of two masses m_1 and m_2 moving with velocities $\vec{v}_1$ and $\vec{v}_2$, respectively, is

$$\vec{J} = \vec{r}_1 \times m_1\vec{v}_1 + \vec{r}_2 \times m_2\vec{v}_2 \tag{4.10}$$

Using equation (4.9) in (4.10), we get

$$\begin{aligned}\vec{J} &= \vec{r}_1 \times m_1\vec{v}_1 - \vec{r}_2 \times m_1\vec{v}_1 \\ &= (\vec{r}_1 - \vec{r}_2) \times m_1\vec{v}_1 \end{aligned} \tag{4.11}$$

We can have

$$m_1\vec{v}_1 = \frac{m_1(m_1+m_2)\vec{v}_1}{(m_1+m_2)} = \frac{m_1}{(m_1+m_2)}\,(m_1\vec{v}_1 + m_2\vec{v}_1) \tag{4.12}$$

Using equation (4.9) in (4.12), we get

$$m_1\vec{v}_1 = \frac{m_1}{(m_1+m_2)}\,(-m_2\vec{v}_2 + m_2\vec{v}_1)$$

$$= \frac{m_1 m_2}{(m_1 + m_2)} (\vec{v}_1 - \vec{v}_2) \tag{4.13}$$

Using equation (4.13) in (4.11), we get

$$\vec{J} = (\vec{r}_1 - \vec{r}_2) \times \frac{m_1 m_2}{(m_1 + m_2)} (\vec{v}_1 - \vec{v}_2) \tag{4.14}$$

Expressing

$$\vec{r} = \vec{r}_1 - \vec{r}_2 \tag{4.15}$$

and differentiating equation (4.15) with respect to time, we get

$$\vec{v} = \vec{v}_1 - \vec{v}_2 \tag{4.16}$$

Using equations (4.15) and (4.16) in (4.14), we get

$$\vec{J} = \vec{r} \times \mu \, \vec{v}$$

Here, $\vec{r}$ is the position vector of mass m_1 relative to that of mass m_2, $\vec{v}$ the velocity of mass m_1 relative to that of mass m_2 and μ the reduced mass.

Energy

Total energy of the system is

$$E = \frac{1}{2} m_1 \vec{v}_1 \,.\, \vec{v}_1 + \frac{1}{2} m_2 \vec{v}_2 \,.\, \vec{v}_2 - \frac{G m_1 m_2}{r}$$

Using $\vec{v}_2 = -m_1 \vec{v}_1 / m_2$ from equation (4.9), we get

$$E = \frac{1}{2} m_1 \vec{v}_1 \,.\, \vec{v}_1 + \frac{1}{2} m_1 \vec{v}_1 \,.\, \frac{m_1 \vec{v}_1}{m_2} - \frac{G m_1 m_2}{r}$$

$$= \frac{1}{2} \left(m_1 + \frac{m_1^2}{m_2} \right) \vec{v}_1 \,.\, \vec{v}_1 - \frac{G m_1 m_2}{r}$$

Using

$$\vec{v}_1 = \frac{m_2}{(m_1 + m_2)} (\vec{v}_1 - \vec{v}_2)$$

from equation (4.14), here, we get

$$E = \frac{1}{2} \left(m_1 + \frac{m_1^2}{m_2} \right) \frac{m_2^2}{(m_1 + m_2)^2} (\vec{v}_1 - \vec{v}_2).(\vec{v}_1 - \vec{v}_2) - \frac{G m_1 m_2}{r}$$

$$= \frac{1}{2} \frac{m_1 m_2}{m_1 + m_2} (\vec{v}_1 - \vec{v}_2).(\vec{v}_1 - \vec{v}_2) - \frac{G m_1 m_2}{r}$$

$$= \frac{1}{2} \mu v^2 - \frac{G m_1 m_2}{r}$$

where $\vec{v} = \vec{v}_1 - \vec{v}_2$ is the velocity of mass m_1 relative to that of the mass m_2 and μ is the reduced mass. It may also be expressed as

$$E = \frac{1}{2} \mu v^2 - \frac{G \mu M}{r}$$

where $M = m_1 + m_2$ is the total mass of the system.

4.7 Kepler laws of planetary motion

After analysing the data about the position and velocity of Mars on various dates, Kepler derived three fundamental laws, which have been found applicable to all the planets revolving around the sun. These laws known as the Kepler's laws for planetary motion may be stated as following:

1. A planet revolves around the sun in an elliptical orbit. This is known as the law of orbit.

2. The line joining one of the foci of the ellipse of orbit and the planet sweeps equal area in equal time. That is, the areal velocity of the planet is constant. This is known as the law of areal velocity.

   As shown in the following figure, a planet moves in equal times from P to P_1 and from P_2 to P_3 and the areas PSP_1 and P_2SP_3 are equal. Obviously, the speed of the planet during the motion from P to P_1 will be faster as compared to that during the motion from P_2 to P_3.

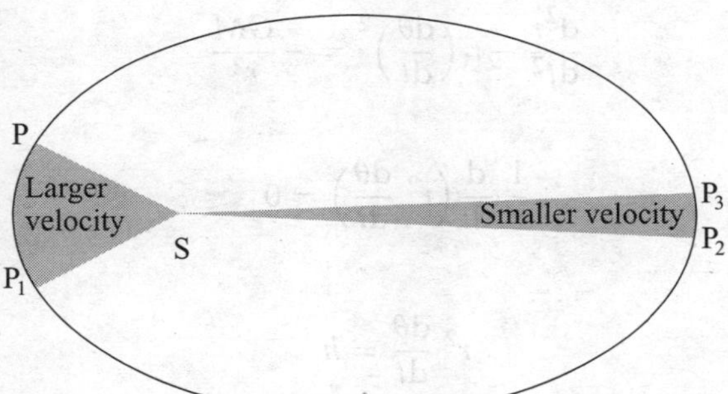

3. For a planet, the square of time period for one revolution around the sun is proportional to the cube of the semi-major axis of the ellipse of the path. This is known as the law of time period.

It has been found interesting to note that though these three laws were first derived from the data for the planet Mars, but they have been found valid for all other planets as well. These three laws can be derived in the following manner:

Let us consider two bodies A and B of masses m_1 and m_2, respectively, revolving around their centre of mass. Suppose in an inertial reference frame, at time t the bodies are situated at position vectors $\vec{r}_1$ and $\vec{r}_2$, respectively. It is well established that such a system of two moving bodies can be treated as equivalent to one body of mass μ, expressed as $\mu = m_1m_2/(m_1+m_2)$, revolving at a distance $r(= \mid \vec{r}_2 - \vec{r}_1 \mid)$ around a stationary mass M, where $M = m_1 + m_2$. Let the velocity of the mass μ be $\vec{v}$. Then, the total energy E and total orbital angular momentum $\vec{L}$ of the system are

$$E = \frac{1}{2}\mu v^2 - G\frac{\mu M}{r} \qquad \text{and} \qquad \vec{J} = \mu \ \vec{r} \times \vec{v}$$

The force of attraction between the two particles of masses μ and M is

$$\vec{F} = -\frac{G\mu}{r^2}\hat{r}$$

where $\hat{r}$ is the unit vector along the radial distance. This is the case of central force. A planet around the sun revolves under the central force. Thus, the revolving particle of mass μ has an acceleration

$$\vec{a} = -\frac{GM}{r^2}\hat{r} \tag{4.17}$$

When we use the polar coordinates (r, θ) with origin at the mass M, this acceleration can be resolved into the radial and transverse components:

$$\vec{a} = \left[\frac{d^2r}{dt^2} - r\left(\frac{d\theta}{dt}\right)^2\right]\hat{r} + \frac{1}{r}\frac{d}{dt}\left(r^2\frac{d\theta}{dt}\right)\hat{\theta} \tag{4.18}$$

Comparison of equations (4.17) and (4.18) gives

$$\frac{d^2r}{dt^2} - r\left(\frac{d\theta}{dt}\right)^2 = -\frac{GM}{r^2} \tag{4.19}$$

and

$$\frac{1}{r}\frac{d}{dt}\left(r^2\frac{d\theta}{dt}\right) = 0 \tag{4.20}$$

Equation (4.20) gives

$$r^2\frac{d\theta}{dt} = h$$

where h is some constant and we have

$$\frac{d\theta}{dt} = \frac{h}{r^2} \tag{4.21}$$

Using equation (4.21) in (4.19), we get

$$\frac{d^2r}{dt^2} - \frac{h^2}{r^3} + \frac{GM}{r^2} = 0 \quad \text{or} \quad \frac{d^2r}{dt^2} + \frac{d}{dr}\left(-\frac{GM}{r} + \frac{h^2}{2r^2}\right) = 0$$

Multiplying this equation by dr/dt, we get

$$\frac{d^2r}{dt^2}\frac{dr}{dt} + \frac{d}{dr}\left(-\frac{GM}{r} + \frac{h^2}{2r^2}\right)\frac{dr}{dt} = 0$$

$$\frac{d}{dt}\left[\frac{1}{2}\left(\frac{dr}{dt}\right)^2\right] + \frac{d}{dt}\left(-\frac{GM}{r} + \frac{h^2}{2r^2}\right) = 0$$

$$\frac{d}{dt}\left[\frac{1}{2}\left(\frac{dr}{dt}\right)^2 - \frac{GM}{r} + \frac{h^2}{2r^2}\right] = 0 \tag{4.22}$$

Integration of equation (4.22) gives

$$\frac{1}{2}\left(\frac{dr}{dt}\right)^2 - \frac{GM}{r} + \frac{h^2}{2r^2} = \frac{E}{\mu} \tag{4.23}$$

where E/μ is a constant. Here, E is the total (kinetic + potential) energy of the system.[2] Equation (4.23) can be rearranged as

$$\frac{dr}{dt} = \sqrt{2\left(\frac{E}{\mu} + \frac{GM}{r} - \frac{h^2}{2r^2}\right)} \tag{4.24}$$

Dividing equation (4.24) by (4.21), we get

$$\frac{dr}{d\theta} = \frac{r^2}{h}\sqrt{2\left(\frac{E}{\mu} + \frac{GM}{r} - \frac{h^2}{2r^2}\right)}$$

or

$$d\theta = \frac{h}{r^2}\frac{dr}{\sqrt{2\left(\frac{E}{\mu} + \frac{GM}{r} - \frac{h^2}{2r^2}\right)}} \tag{4.25}$$

This equation can be solved by using the relation

$$r = \frac{1}{u} \qquad \text{so that} \qquad dr = -\frac{1}{u^2}\,du$$

and equation (4.25) gives

$$d\theta = -\frac{du}{\sqrt{\frac{2E}{\mu h^2} + \frac{2GMu}{h^2} - u^2}}$$

$$= -\frac{du}{\sqrt{\left(\frac{2E}{\mu h^2} + \frac{G^2M^2}{h^4}\right) - \left(u - \frac{GM}{h^2}\right)^2}}$$

On integration, we get

$$\theta = \cos^{-1}\left(\frac{u - \frac{GM}{h^2}}{\sqrt{\frac{2E}{\mu h^2} + \frac{G^2M^2}{h^4}}}\right) + \theta_0$$

[2]In the polar coordinates, radial and transverse components of velocity are, respectively,

$$v_r = \frac{dr}{dt} \qquad \text{and} \qquad v_\theta = r\,\frac{d\theta}{dt}$$

Thus, total (kinetic + potential) energy is

$$E = \frac{1}{2}\mu\left[\left(\frac{dr}{dt}\right)^2 + r^2\left(\frac{d\theta}{dt}\right)^2\right] + \left(-\frac{GM\mu}{r}\right)$$

Using equation (4.21) we get

$$\frac{E}{\mu} = \frac{1}{2}\left(\frac{dr}{dt}\right)^2 + \frac{h^2}{2r^2} - \frac{GM}{r}$$

where θ_0 is the constant of integration. This equation can be rearranged as

$$u - \frac{GM}{h^2} = \left[\sqrt{\frac{2E}{\mu h^2} + \frac{G^2M^2}{h^4}}\right] \cos(\theta - \theta_0)$$

$$\frac{1}{r} = \frac{GM}{h^2} + \frac{GM}{h^2}\left[\sqrt{1 + \frac{2h^2E}{\mu G^2M^2}}\right] \cos(\theta - \theta_0)$$

$$= \frac{GM}{h^2}\left\{1 + \left[\sqrt{1 + \frac{2h^2E}{\mu G^2M^2}}\right] \cos(\theta - \theta_0)\right\}$$

Thus, we have

$$r = \frac{h^2/GM}{1 + \left[\sqrt{1 + \frac{2h^2E}{\mu G^2M^2}}\right] \cos(\theta - \theta_0)} = \frac{a(1 - e^2)}{1 + e\cos(\theta - \theta_0)}$$

This is equation of an ellipse with semi-major axis a and eccentricity e, showing that the path of the motion under the central force is elliptical. This is the Kepler's first law of motion of a particle under the gravitational force. Here, θ_0 is value of θ when r is minimum, that is at the periastron position

$$e = \sqrt{1 + \frac{2h^2E}{\mu G^2M^2}} \tag{4.26}$$

and

$$a(1 - e^2) = \frac{h^2}{GM} \tag{4.27}$$

Using equation (4.26) in (4.27) we get

$$a = -\frac{GM\mu}{2E}$$

It shows that the semi-major axis of the orbit is inversely proportional to the total energy of the system.

Suppose the body of mass μ moving in the elliptical path is at a position P at time t (Figure 4.8). After a time interval dt, the body reaches at Q. Thus, the area traversed by the line, joining the focus O of the ellipse to the body, in time dt is

$$\text{Area} = \frac{1}{2}\,\text{OP} \times \text{PQ} = \frac{1}{2}\,r^2\,\mathrm{d}\theta$$

Hence, the areal velocity is

$$\text{Areal velocity} = \frac{\text{area traversed}}{\text{time interval}} = \frac{1}{2}\,r^2\frac{\mathrm{d}\theta}{\mathrm{d}t} = \frac{h}{2} \tag{4.28}$$

where equation (4.28) is used. This shows that for a particle moving in an elliptical path, the areal velocity is constant. This is the Kepler's second law of motion.

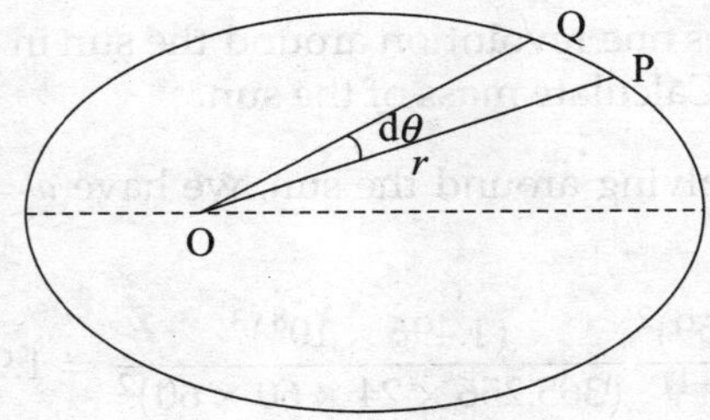

Figure 4.8: Body of mass μ revolving in the elliptical path around a stationary body of mass M is at a position P at time t. After a time interval dt, the body reaches at Q.

Area of an ellipse with semi-major axis a and semi-minor axis b is πab. If e is the eccentricity of the ellipse, we have $b = a\sqrt{1-e^2}$. If T is the time period of motion, we have

$$\frac{h}{2} = \frac{\pi ab}{T} = \frac{\pi a^2\sqrt{1-e^2}}{T} \tag{4.29}$$

Using equation (4.27) in (4.29), we have

$$\frac{h}{2} = \frac{\pi a^2}{T}\sqrt{\frac{h^2}{GMa}}$$

Thus, we have

$$\frac{a^3}{T^2} = \frac{GM}{4\pi^2} = \frac{G(m_1+m_2)}{4\pi^2} \tag{4.30}$$

This shows that square of the time period is proportional to the cube of the semi-major axis. This is the Kepler's third law of motion.

Case of the solar system

For the solar system, mass m_2 of a planet is negligible in comparison to the mass $m_1 = M_\odot$ of the sun ($m_2 << M_\odot$). Hence,

$$\frac{a^3}{T^2} = \frac{GM_\odot}{4\pi^2} = \text{constant}$$

Thus, from the Kepler's third law of motion, we have

$$M_\odot = \frac{4\pi^2}{G}\frac{a^3}{T^2}$$

For a circular orbit with radius a we have $T = 2\pi a/v$, where v is the velocity of revolution and thus, we get

$$M_\odot = \frac{4\pi^2}{G}\frac{a^3}{(2\pi a/v)^2} = \frac{av^2}{G}$$

It shows that with the increase of the radius of the orbit, the orbital velocity of the planet decreases in such a manner that av^2 remains constant.

Exercise 4.8: Earth completes one revolution around the sun in 365.256 days at an average distance of 1.496×10^8 km. Calculate mass of the sun.

Solution: For the earth revolving around the sun, we have $a = 1.496 \times 10^8$ km and $T = 365.256$ days and thus

$$M_\odot = \frac{4 \times (3.14159)^2}{6.67 \times 10^{-11}} \frac{(1.496 \times 10^8)^3}{(365.256 \times 24 \times 60 \times 60)^2} = 1.99 \times 10^{30} \text{ kg}$$

Case of the earth-moon system

For the earth-moon system, the mass m_2 of the moon is not negligible in comparison to the mass m_1 of the earth. Hence, equation (4.30) gives the sum of the masses $(m_1 + m_2)$. For the moon revolving around the earth, we have $a = 384404$ km and $T = 27.322$ days and we have

$$m_1 + m_2 = \frac{4 \times (3.14159)^2}{6.67 \times 10^{-11}} \frac{(384404 \times 10^3)^3}{(27.322 \times 24 \times 60 \times 60)^2} = 6.033 \times 10^{24} \text{ kg}$$

After subtracting the mass of the moon, we can find out the mass of the earth.

Exercise 4.9: A planet is revolving around the sun in an elliptical orbit. If the minimum and maximum distances of the planet from the sun are r_{min} and r_{max}, respectively. Show that the eccentricity of the orbit is

$$e = \frac{r_{max} - r_{min}}{r_{max} + r_{min}}$$

Solution: Using plane polar coordinates (r, θ) in a plane, the equation of elliptical orbit is

$$r = \frac{a(1-e^2)}{1 + e\cos\theta}$$

where e and a denote eccentricity and semi-major axis, respectively, of the ellipse. The distance of planet from the sun is minimum when $\theta = 0$ and we have

$$r_{min} = \frac{a(1-e^2)}{1+e} = a(1-e)$$

The distance of planet from the sun is maximum when $\theta = \pi$ and we have

$$r_{max} = \frac{a(1-e^2)}{1-e} = a(1+e)$$

Thus, we have

$$r_{max} + r_{min} = 2a \qquad \text{and} \qquad r_{max} - r_{min} = 2ae$$

It gives

$$e = \frac{r_{max} - r_{min}}{r_{max} + r_{min}}$$

Exercise 4.10: A planet of mass is revolving around the sun in elliptical orbit with eccentricity $e = \sqrt{1 + \alpha E}$ and semi-major axis $a = -\beta/E$, where E is total energy, and α and β

are constants. Using Kepler's second law of planetary motion, derive the Kepler's third law of planetary motion.

Solution: For an ellipse with semi-major axis a, semi-minor axis b and eccentricity e, we have

$$b = a\sqrt{1 - e^2}$$

and area of ellipse is πab. According to the Kepler's second law of planetary motion, the areal velocity is constant, say v_A. If T denotes the time period of planetary motion, we have

$$v_A = \frac{\pi ab}{T} = \frac{\pi a^2\sqrt{1-e^2}}{T}$$

Using the value of e here, we get

$$v_A T = \pi a^2\sqrt{1-(1+\alpha E)} \qquad \text{or} \qquad v_A^2 T^2 = -\pi^2 a^4 \alpha E$$

Using the value of E, we get

$$v_A^2 T^2 = \pi^2 a^3 \alpha\beta \qquad \text{or} \qquad T^2 \propto a^3$$

It gives the Kepler's third law of planetary motion.

4.8 Escape velocity

Escape velocity is defined as the velocity that an object requires to escape from the surface of a heavenly body such as a planet or satellite. Let us consider a case of escape from a surface of the earth of mass M and radius R. Suppose an object of mass m is thrown with velocity v from the surface of the earth. When the object is at a distance r from the centre of the earth, the force of attraction between the earth and object is

$$F = -G\frac{Mm}{r^2}$$

The potential energy V is expressed as

$$F = -\frac{\mathrm{d}V}{\mathrm{d}r} \qquad \text{or} \qquad V = -\int F\,\mathrm{d}r + C$$

where C is the constant of integration. Using the expression for F, we get

$$V = -G\frac{Mm}{r} + C$$

Since $V = 0$, when $r = \infty$, we have $C = 0$ and therefore

$$V = -G\frac{Mm}{r}$$

When the object is on the surface of the earth, the potential energy is

$$V = -G\frac{Mm}{R}$$

Thus, in order to send the object out of the earth's gravitational field at least this much amount of kinetic energy be given to the object. The minimum velocity v_e given to the object so that it will go away from the earth's gravitational field is known as the escape velocity. Thus, we have

$$\frac{1}{2}\,mv_e^2 = G\,\frac{Mm}{R} \qquad \text{or} \qquad v_e = \sqrt{\frac{2GM}{R}} \tag{4.31}$$

It is interesting to note that the escape velocity does not depend on the mass of the object. If g denotes the acceleration due to gravity of the earth, we have

$$g = \sqrt{\frac{GM}{R^2}} \tag{4.32}$$

From equations (4.31) and (4.32), we have

$$v_e = \sqrt{2gR}$$

Using the values $g = 9.8\ \text{m/s}^2$ and $R = 6.4 \times 10^6$ m, we have

$$v_e = \sqrt{2 \times 9.8 \times 6.4 \times 10^6} = 11200\ \text{m/s}$$

Thus, the escape velocity of the earth is 11.2 km/s.

Exercise 4.11: Mass and radius of moon are 7.34×10^{22} kg and 1.74×10^6 m, respectively. Calculate the escape velocity for the moon. (Given $G = 6.67 \times 10^{-11}$ N-m^2/kg^2.)

Solution: The escape velocity is

$$v_e = \sqrt{\frac{2GM}{R}} = \sqrt{\frac{2 \times 6.67 \times 10^{-11} \times 7.34 \times 10^{22}}{1.74 \times 10^6}} = 2372\ \text{m/s}$$

The escape velocity of the moon is 2.4 km/s. It is much smaller than that for the earth. Because of this low value of escape velocity, the moon does not have any atmosphere, since all the material from the surface of the moon escape away.

4.8.1 Existence of atmosphere around a heavenly body

For a heavenly body of mass M, radius R and acceleration due to gravity g, the escape velocity is

$$v_e = \sqrt{\frac{2GM}{R}} = \sqrt{2gR}$$

That is, an object having velocity more than v_e will escape from the surface of the heavenly body. Suppose T is the temperature of atmosphere around the body. The gas of molecular mass m in the atmosphere will have velocity expressed as

$$\frac{1}{2}\,mv^2 = \frac{3}{2}\,kT \qquad \text{or} \qquad v = \sqrt{\frac{3kT}{m}}$$

Thus, if $v > v_e$, the molecules will escape from the atmosphere of the heavenly body. Consequently, the heavenly body will not have atmosphere. For example, the satellite moon has no atmosphere.

However, at a given temperature T, all the molecules do not have the same velocity v. Some molecules may have large velocity and some may have small velocity. The molecules will follow the Maxwellian distribution of velocity and the average velocity is

$$\bar{v} = \sqrt{\frac{8kT}{\pi m}}$$

Following the theory of J.H. Jeans, if a gas is to remain on heavenly body for a period of 1 billion years, the mean speed $\bar{v}$ should be at least $v_e/5$. Thus, for the existence of atmosphere around a heavenly body, we have a condition

$$\sqrt{\frac{8kT}{\pi m}} \le \frac{v_e}{5} \qquad \text{or} \qquad \frac{8kT}{\pi m} \le \frac{2GM}{25R}$$

The minimum molecular mass m that can remain in the atmosphere of a heavenly body is

$$m = \frac{100kTR}{\pi GM}$$

For $T = 300$ K on the earth, we have

$$m = \frac{100 \times 1.38 \times 10^{-23} \times 300 \times 6.4 \times 10^{6}}{3.14 \times 6.67 \times 10^{-11} \times 6 \times 10^{24}} = 2.10 \times 10^{-27}\ \text{kg}$$

Since 1 amu = 1.66×10^{-27} kg, we have $m = 1.27$ amu. Thus, there is no possibility of hydrogen ($m = 1$) to exist in the earth's atmosphere. However, in the upper regions of atmosphere, the temperature is very low and therefore, atomic hydrogen can exist in the earth's atmosphere.

Let us look for the moon with atmospheric temperature $T = 400$ K. We have

$$m = \frac{100 \times 1.38 \times 10^{-23} \times 400 \times 1.7 \times 10^{6}}{3.14 \times 6.67 \times 10^{-11} \times 7.3 \times 10^{22}}$$

$$= 6.14 \times 10^{-26}\ \text{kg} = 37\ \text{amu}$$

It shows that the gases nitrogen (28) and oxygen (32), which are the predominant constituents of earth's atmosphere, cannot be present on the moon.

4.9 Motion of a satellite in a circular orbit

Satellite is an object which revolves around a planet. Moon is the natural satellite of the earth. Besides the moon, there are artificial satellites revolving around the earth of mass M. Although the orbit of a satellite is elliptical, but as an approximation, we shall take it as a circle (Figure 4.9). Suppose a satellite of mass m is revolving around the earth at a distance r from the centre of the earth. The force of acceleration between the earth and satellite is

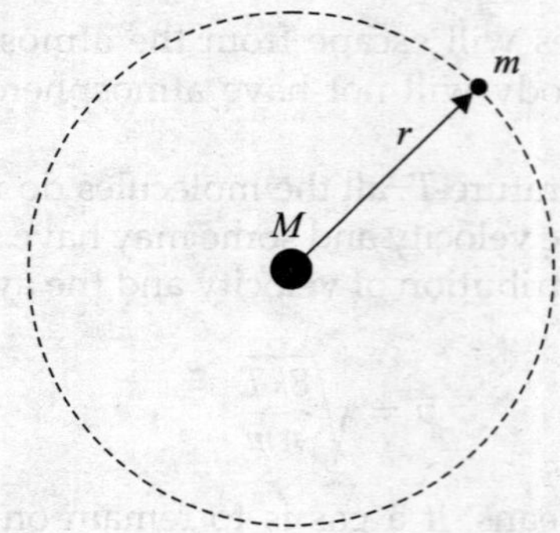

Figure 4.9: A satellite of mass m revolving around the earth of mass M in a circular orbit of radius r.

$$F = -\frac{GMm}{r^2}$$

This force is balanced by the centrifugal force acting on the satellite due to its angular motion. If v is the velocity of the satellite, the centrifugal force acting in the outward direction is

$$F' = \frac{mv^2}{r}$$

In the equilibrium position, we have

$$\frac{GMm}{r^2} = \frac{mv^2}{r} \quad \text{or} \quad v = \sqrt{\frac{GM}{r}}$$

If R is the radius of the earth and h the altitude of the satellite above the surface of the earth, we have $r = R + h$, and therefore orbital velocity

$$v_o = \sqrt{\frac{GM}{R+h}}$$

Near the surface of the earth, we have $h << R$, therefore

$$v_o = \sqrt{\frac{GM}{R}}$$

Using the relation

$$g = \sqrt{\frac{GM}{R^2}} \quad \text{we get} \quad v_o = \sqrt{gR}$$

Thus, we have $v_e = \sqrt{2}\, v_o$. Using $g = 9.8\ \text{m/s}^2$ and $R = 6.4 \times 10^6$ m, we have

$$v_o = \sqrt{9.8 \times 6.4 \times 10^6} = 7.9\ \text{m/s}$$

The period of revolution T of the satellite around the planet is

$$T = \frac{2\pi r}{v_o} = \frac{2\pi}{\sqrt{GM}} r^{3/2}$$

4.9.1 Geostationary satellite

For a geostationary satellite, the time period of revolution is equal to the time period of spin motion of the earth. It moves from west towards east. The earth completes one rotation about its own axis in 24 hours.[3] The distance r of the satellite from the centre of the earth is

$$r = \left(\frac{GM}{4\pi^2}\right)^{1/3} T^{2/3}$$

Using $T = 24$ hr = 86400 s, $G = 6.67 \times 10^{-11}$ m^3/kg-s^2 and mass of the earth $M = 6 \times 10^{24}$ kg, we have

$$r = \left(\frac{6.67 \times 10^{-11} \times 6 \times 10^{24}}{4\pi^2}\right)^{1/3} (86400)^{2/3} = 42.3 \times 10^6 \text{ m}$$

Thus, the distance of the satellite from the centre of the earth is about 42.3×10^6 m. For the radius of the earth $R = 6.4 \times 10^6$ m, the distance of the satellite from the earth's surface is about 35.9×10^6 m.

Multiple-choice questions

1. Which of the following forces is not central force?

   (i) Force between two charges (ii) Lorentz force

   (iii) Force between two magnetic poles (iv) Gravitational force

   Ans. (ii)

2. A spherical shell has mass M and radius R. The gravitational potential at its centre is

   (i) Zero (ii) $-GMm/R$

   (iii) $-GMm/R^2$ (iv) $-GMmR$

   Ans. (iii)

3. A spherical shell has mass M and radius R. The gravitational field at its centre is

   (i) Zero (ii) $-GMm/R$

   (iii) $-GMm/R^2$ (iv) $-GMmR$

   Ans. (i)

4. Gravitational self-energy of a sphere of mass M and radius R is

   (i) Zero (ii) $-3GM^2/5R$

   (iii) $-GM/R^2$ (iv) $-3GM^2/5R^2$

   Ans. (ii)

[3] Actual time period is 23 hr, 56 min, 4.1 s.

5. The reduced mass μ of a system of three masses m_1, m_2, m_3 is

(i) $\frac{1}{\mu} = \frac{1}{m_1} + \frac{1}{m_2+m_3}$ (ii) $\frac{1}{\mu} = \frac{1}{m_1+m_2} + \frac{1}{m_3}$

(iii) $\frac{1}{\mu} = \frac{1}{m_1} + \frac{1}{m_2} + \frac{1}{m_3}$ (iv) $\frac{1}{\mu} = \frac{1}{m_1+m_3} + \frac{1}{m_2}$

Ans. (iii)

6. Mass of hydrogen atom is m. The reduced mass of hydrogen molecule is

(i) $m/2$ (ii) $2m$

(iii) $m/4$ (iv) $4m$

Ans. (i)

7. A planet revolves around the sun in an elliptical path with major axis $2a$ and minor axis $2b$. The time of revolution of the planet is

(i) $T \propto a^{3/2}$ (ii) $T \propto a$

(iii) $T \propto b^{3/2}$ (iv) $T \propto b$

Ans. (i)

8. The eccentricity $e = 0$ for

(i) Circle (ii) Ellipse

(iii) Parabola (iv) Hyperbola

Ans. (i)

9. The eccentricity $e = 1$ for

(i) Circle (ii) Ellipse

(iii) Parabola (iv) Hyperbola

Ans. (iii)

10. The eccentricity $e > 1$ for

(i) Circle (ii) Ellipse

(iii) Parabola (iv) Hyperbola

Ans. (iv)

11. The eccentricity $0 < e < 1$ for

(i) Circle (ii) Ellipse

(iii) Parabola (iv) Hyperbola

Ans. (ii)

12. In the context of a geostationary satellite, which of the following statements is correct?

(i) A geostationary satellite moves from east towards west.

(ii) A geostationary satellite moves from west towards east.

(iii) Revolution period for a geostationary satellite is 12 hr.

(iv) A geostationary satellite revolves around the sun.

Ans. (ii)

13. In the context of a geostationary satellite, which of the following statements is correct?

(i) A geostationary satellite moves from east towards west.

(ii) Revolution period for a geostationary satellite is 12 hr.

(iii) Revolution period for a geostationary satellite is 24 hr.

(iv) A geostationary satellite revolves around the sun.

Ans. (iii)

Exercises

1. Express central force. Show that for a central force, the torque is zero and the angular momentum of a body moving under the central force is conserved.
2. A planet of mass m is revolving around the sun having mass M. The separation between their centres is r. Show that the angular momentum of revolution is conserved.
3. A satellite of mass m is revolving around its planet having mass M. The separation between their centres is r. Show that the angular momentum of revolution is conserved.
4. An electron carrying a charge q_1 is revolving around a nucleus carrying a charge q_2. The separation between the electron and nucleus is r. Show that the angular momentum of revolution is conserved.
5. Derive an expression for potential due to a spherical shell at a point situated: (i) outside the shell and (ii) inside the shell.
6. Derive an expression for potential due to a solid sphere at a point situated: (i) outside the shell and (ii) inside the shell.
7. A point mass m is placed at the centre of a solid sphere of mass M and radius R. Show that the potential between the point mass and the sphere is
$$V = -\frac{3GmM}{2R}$$
8. A thick spherical shell of mass M has inner radius R_1 and outer radius R_2. Calculate the potential due to the shell and a point mass m placed at the centre of the shell.
9. A galaxy consists of n stars, each of radius R and mass M. Obtain self-energy of the galaxy.
10. Show that gravitational self-energy of a uniform sphere of mass M and radius R is
$$-\frac{3}{5}\frac{GM^2}{R}$$

11. Define reduced mass of two masses m_1 and m_2. What will happen when: (i) $m_1 << m_2$, and (ii) $m_1 << m_2$?

12. Two interacting masses m_1 and m_2 are orbiting about the centre of mass. Show how this two-body problem can be reduced to one-body problem.

13. When a two-body problem is reduced to one-body problem, how the angular momentum and energy of one-body problem are expressed?

14. Show that angular momentum of a system of two masses m_1 and m_2 can be expressed as

$$\vec{J} = \vec{r} \times \mu \vec{v}$$

where $\vec{r}$ is the position vector of mass m_1 relative to that of mass m_2, $\vec{v}$ is the velocity of mass m_1 relative to that of mass m_2 and μ is the reduced mass.

15. Show that total energy of a system of two masses m_1 and m_2 can be expressed as

$$E = \frac{1}{2}\mu v^2 - \frac{Gm_1m_2}{r}$$

where $\vec{r}$ is the position vector of mass m_1 relative to that of mass m_2, $\vec{v}$ is the velocity of mass m_1 relative to that of mass m_2 and μ is the reduced mass.

16. Describe Kepler laws of planetary motion and obtain the equation of orbit.

17. Write Kepler laws of planetary motion. Show that the path of a planet revolving around the sun is an ellipse.

18. Show that for motion of a planet revolving around the sun, the semi-major axis of the elliptical path is inversely proportional to the energy of the planet.

19. When we use polar coordinates (r, θ) for motion in a plane, the acceleration is expressed as

$$\vec{a} = \left[\frac{d^2r}{dt^2} - r\left(\frac{d\theta}{dt}\right)^2\right]\hat{r} + \frac{1}{r}\frac{d}{dt}\left(r^2\frac{d\theta}{dt}\right)\hat{\theta}$$

For motion of a planet around the sun, show that the areal velocity is constant.

20. Derive expression for escape velocity for a planet. Show that the escape velocity does not depend on the mass of the object.

21. Describe motion of a satellite around the earth in a circular orbit.

22. What is a geostationary satellite? Calculate the altitude of a geostationary satellite revolving in a circular orbit around the earth.

23. Write notes on the following:

(i) Central force

(ii) Inverse-square-law force

(iii) Gravitational self-energy

(iv) Gravitational self-energy of a uniform sphere

(v) Reduction of two-body problem into one-body problem

(vi) Kepler laws of planetary motion

(vii) Escape velocity

(viii) Geostationary satellite

CHAPTER 5

HARMONIC OSCILLATOR

A motion in which a body moves to and fro about a position, generally called the mean position, is said to be harmonic. A harmonic oscillator is one of the good examples in mechanics.

5.1 Harmonic motion

When the position of a body repeats after a regular interval of time, the motion is said to be harmonic or oscillatory. Example of such motion may be the motion of a body moving along a circular path with constant speed. For a harmonic motion, we define the following parameters.

Cycle

When a body executing harmonic motion reaches again at the same position and in the same phase, the body is said to complete one cycle.

Time period

The time period of a harmonic motion is the time required to complete one cycle. It is generally denoted by T and is expressed in seconds.

Frequency

The frequency of a harmonic motion is the number of cycles completed in 1 s. It is generally denoted by ν and is expressed in the units of per second or Hertz. Obviously, the relation between time period T and frequency ν is

$$\nu = \frac{1}{T} \qquad \text{or} \qquad T = \frac{1}{\nu}$$

5.1.1 Simple harmonic motion

When the harmonic motion of a body is along a straight line, the motion is said to be simple harmonic motion. Example of such motion may be motion of a pendulum with small amplitude under the gravitational force and vibrational motion of a diatomic molecule. In a simple harmonic motion, the force acting on the moving body is always directed towards the equilibrium (mean) position of the motion and is proportional to the distance of the body from the equilibrium position. This force is generally known as the restoring force, as it is always towards the equilibrium position.

Amplitude

In case of simple harmonic motion, we have one more parameter, known as the amplitude. It is the maximum distance from the equilibrium position moved by the body under simple harmonic motion. If we denote it by A, then the body can move up to a distance A on both sides of the equilibrium position.

5.2 Mathematical treatment of simple harmonic motion

Let us consider a particle (body) P of mass m executing simple harmonic motion. The motion is along a straight line. Suppose at any instant t, the distance of the particle from the mean (equilibrium) position O is x. The restoring force F acting on the particle will be proportional to the distance x and we have

$$F \propto -x \qquad \text{or} \qquad F = -kx$$

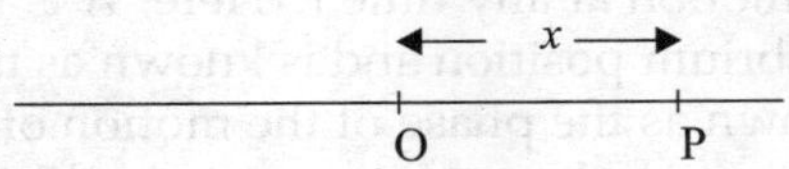

Here, the negative sign shows that the restoring force is always towards the equilibrium position and the proportionality constant k is known as the force constant. When $\mathrm{d}^2x/\mathrm{d}t^2$ is the acceleration of particle of mass m, we have

$$m\,\frac{\mathrm{d}^2x}{\mathrm{d}t^2} = -kx \qquad \text{or} \qquad \frac{\mathrm{d}^2x}{\mathrm{d}t^2} + \omega^2 x = 0 \tag{5.1}$$

where $\omega = \sqrt{k/m}$ is known as the angular frequency. Equation (5.1) is known as the differential equation of simple harmonic motion. In order to get solution of equation (5.1), let us assume

$$x = C\,\mathrm{e}^{\alpha t} \tag{5.2}$$

where C and α are constants. On differentiation of equation (5.2), we have

$$\frac{\mathrm{d}x}{\mathrm{d}t} = C\alpha\,\mathrm{e}^{\alpha t} \qquad \text{and} \qquad \frac{\mathrm{d}^2x}{\mathrm{d}t^2} = C\alpha^2\,\mathrm{e}^{\alpha t} \tag{5.3}$$

Using equations (5.2) and (5.3) in (5.1), we get

$$C\,\mathrm{e}^{\alpha t}(\alpha^2 + \omega^2) = 0$$

Since $C\,e^{\alpha t} \neq 0$, we have

$$\alpha^2 + \omega^2 = 0 \qquad \text{or} \qquad \alpha = \pm i\omega$$

where $i = \sqrt{-1}$ is the imaginary number. Since α has two values, possible solutions of equation (5.1) are

$$x = C\,e^{i\omega t} \qquad \text{and} \qquad x = C\,e^{-i\omega t}$$

A general solution of equation (5.1) is a linear combination of these two solutions and we have

$$x = C_1\,e^{i\omega t} + C_2\,e^{-i\omega t}$$

where C_1 and C_2 are arbitrary constants. This equation can be expressed as

$$x = C_1(\cos\omega t + i\sin\omega t) + C_2(\cos\omega t - i\sin\omega t)$$

$$= (C_1 + C_2)\cos\omega t + i(C_1 - C_2)\sin\omega t$$

Using $(C_1 + C_2) = A\sin\delta$ and $i(C_1 - C_2) = A\cos\delta$, we have

$$x = A\sin\delta\,\cos\omega t + A\cos\delta\,\sin\omega t = A\sin(\omega t + \delta) \tag{5.4}$$

This is the required solution of equation (5.1) and gives the displacement of the particle executing simple harmonic motion at any time t. Here, A is the maximum displacement of the particle from its equilibrium position and is known as the amplitude of the motion. The quantity $(\omega t + \delta)$ is known as the phase of the motion of the particle. At time $t = 0$, the phase δ is known as the initial phase of the motion. When the particle starts motion from its mean position, the value of δ is zero. But, when the particle starts motion from any of the extreme positions, the value of δ is $\pi/2$. Let us see what happens when we replace t by $(t + 2\pi/\omega)$ in equation (5.4). We have

$$x = A\sin(\omega[t + 2\pi/\omega] + \delta)$$

$$= A\sin(\omega t + 2\pi + \delta) = A\sin(\omega t + \delta) \tag{5.5}$$

It shows that the motion of the particle is repeated after a time interval of $(2\pi/\omega)$. Thus, the time period T of the motion is

$$T = \frac{2\pi}{\omega}$$

Using here the value of ω, we have the time period

$$T = 2\pi\sqrt{\frac{m}{k}}$$

The frequency ν of simple harmonic motion is

$$\nu = \frac{1}{T} = \frac{1}{2\pi}\sqrt{\frac{k}{m}}$$

Exercise 5.1: The displacement x of a particle executing simple harmonic motion is expressed as $x = A\sin(\omega t + \delta)$. Show that the time period of the motion is $(2\pi/\omega)$.

Solution: We know that the time period of simple harmonic motion is the time required to complete one cycle. The displacement x of particle from its mean position at time t is

$$x = A\sin(\omega t + \delta)$$

After a time interval $(2\pi/\omega)$, the time becomes $t + (2\pi/\omega)$. At this time, the displacement is

$$x' = A\sin(\omega[t + 2\pi/\omega] + \delta)$$

$$= A\sin(\omega t + 2\pi + \delta) = A\sin(\omega t + \delta)$$

This $x' = x$ shows that after a time interval of $(2\pi/\omega)$, the particle is at the same position. Thus, the time period of simple harmonic motion is $(2\pi/\omega)$.

5.2.1 Velocity and acceleration of the particle

Equation (5.5) for displacement of the particle executing simple harmonic motion can be used fo: finding out the velocity and acceleration of the particle. The velocity v of the particle executing the simple harmonic motion is

$$v = \frac{dx}{dt} = A\omega\cos(\omega t + \delta) = A\omega\sqrt{1 - \sin^2(\omega t + \delta)}$$

Using here the value of $\sin(\omega t + \delta)$ from equation (5.5), we get

$$v = \omega\sqrt{A^2 - x^2}$$

This expression gives the velocity v of the particle at a position x. It shows that the velocity is maximum at the mean position $(x = 0)$ and is expressed as

$$v_m = A\omega = \frac{A2\pi}{T}$$

Further, the velocity v is zero when the particle is at the extreme $(x = A)$. The acceleration a of the particle is

$$a = \frac{d^2x}{dt^2} = -A\omega^2\sin(\omega t + \delta) = -\omega^2 x$$

This expression gives the acceleration a of the particle at a position x. It shows that the acceleration at the mean position $(x = 0)$ is zero. Further, acceleration is maximum at each of the extremes $(x = A)$. That is

$$|a_m| = \omega^2 A$$

5.2.2 Energy of a harmonic oscillator

The energy of a harmonic oscillator is the sum of its kinetic energy and the potential energy. For one-dimensional motion, the relation between the acting force F and the potential energy V is expressed as

$$F = -\frac{dV}{dx}$$

The potential energy is thus

$$dV = -F\,dx = kx\,dx$$

Here, we have used the expression for F. On integration, we have

$$V = \frac{1}{2}kx^2 + C$$

where C is the constant of integration. At the mean position, the restoring force is zero, and therefore at $x = 0$, the potential energy is zero. It gives the constant of integration $C = 0$. The potential energy of the particle is therefore

$$V = \frac{1}{2}kx^2$$

It shows that the potential energy is maximum at the extremes of the position. It changes from the maximum value to zero when the particle moves from any of the extreme positions to the mean position. Using the expressions for the force constant k and the displacement x, we have

$$V = \frac{1}{2}m\omega^2 A^2 \sin^2(\omega t + \delta) \tag{5.6}$$

The kinetic energy of the particle is

$$K = \frac{1}{2}mv^2$$

Using the expression for v, we have

$$K = \frac{1}{2}m\omega^2 A^2 \cos^2(\omega t + \delta) \tag{5.7}$$

The kinetic energy can also be expressed as

$$K = \frac{1}{2}m\omega^2[A^2 - A^2\sin^2(\omega t + \delta)] = \frac{1}{2}m\omega^2(A^2 - x^2)$$

It shows that the kinetic energy is maximum at the mean position and it becomes zero at the extremes. The total energy E of the particle is

$$E = K + V$$

Using equations (5.6) and (5.7), we get

$$E = \frac{1}{2}m\omega^2 A^2\cos^2(\omega t + \delta) + \frac{1}{2}m\omega^2 A^2\sin^2(\omega t + \delta) = \frac{1}{2}m\omega^2 A^2$$

Using $T = 2\pi/\omega$, total energy can be expressed as

$$E = \frac{2m\pi^2 A^2}{T^2}$$

Since total energy of the particle does not depend on its position, the energy E of the particle executing the simple harmonic motion is constant. It shows that during simple harmonic motion, the kinetic energy changes into the potential energy and vice versa.

Exercise 5.2: The displacement x of a particle of mass m executing simple harmonic motion with force constant k is expressed as $x = A\sin(\omega t + \delta)$. Show that the average potential energy of the particle is equal to its average kinetic energy.

Solution: The potential energy V and kinetic energy K of the particle executing simple harmonic motion are

$$V = \frac{1}{2}m\omega^2 A^2 \sin^2(\omega t + \delta) \qquad \text{and} \qquad K = \frac{1}{2}m\omega^2 A^2 \cos^2(\omega t + \delta)$$

Obviously, the potential and kinetic energies vary with time and they are never negative. We can compute average of them over a time period $T = 2\pi/\omega$. The average potential energy over the time period is

$$V = \frac{1}{T}\int_0^T V\,dt$$

Using expression for V, we get

$$V = \frac{1}{2T}\int_0^T m\omega^2 A^2 \sin^2(\omega t + \delta)\,dt$$

$$= \frac{m\omega^2 A^2}{4T}\int_0^T [1 - \cos 2(\omega t + \delta)]\,dt$$

$$= \frac{m\omega^2 A^2}{4T}\left[t - \frac{\sin 2(\omega t + \delta)}{2\omega}\right]_0^T$$

$$= \frac{m\omega^2 A^2}{4T}\left[T - \frac{\sin 2(\omega T + \delta)}{2\omega} + \frac{\sin 2\delta}{2\omega}\right] = \frac{m\omega^2 A^2}{4}$$

Here, we used $T = 2\pi/\omega$. The average kinetic energy of the particle is

$$K = \frac{1}{T}\int_0^T K\,dt$$

Using expression for K, we get

$$K = \frac{1}{2T}\int_0^T m\omega^2 A^2 \cos^2(\omega t + \delta)\,dt$$

$$= \frac{m\omega^2 A^2}{4T} \int_0^T [1 + \cos 2(\omega t + \delta)]\, dt$$

$$= \frac{m\omega^2 A^2}{4T} \left[t + \frac{\sin 2(\omega t + \delta)}{2\omega} \right]_0^T$$

$$= \frac{m\omega^2 A^2}{4T} \left[T + \frac{\sin 2(\omega T + \delta)}{2\omega} - \frac{\sin 2\delta}{2\omega} \right] = \frac{m\omega^2 A^2}{4}$$

Here, we used $T = 2\pi/\omega$. It shows that the average potential energy of a particle executing simple harmonic motion is equal to the average kinetic energy of the particle.

Exercise 5.3: A particle of mass 100 g is executing simple harmonic motion with time period of 2π s. The displacement of the particle from its equilibrium position is 80 cm at time $\pi/4$ s. Calculate amplitude of oscillations and total energy of the particle.

Solution: Time period $T = 2\pi$ s. Mass $m = 100$ g = 0.1 kg. Displacement $x = 80$ cm = 0.8 m at time $t = \pi/4$ s. For simple harmonic motion, the displacement is

$$x = A \sin(\omega t) = A \sin\left(\frac{2\pi}{T} t\right)$$

Thus, we have

$$0.8 = A \sin\left(\frac{2\pi}{2\pi} \frac{\pi}{4}\right) = \frac{A}{\sqrt{2}} \qquad \text{or} \qquad A = 0.8\sqrt{2}$$

The amplitude of oscillations is $A = 0.8\sqrt{2} = 1.13$ m. Total energy of the particle is

$$E = \frac{2m\pi^2 A^2}{T^2} = \frac{2 \times 0.1 \times \pi^2 \times 0.8^2 \times 2}{4\pi^2} = 0.064\ \text{J}$$

Exercise 5.4: A particle executing simple harmonic motion with a time period of 2π s has total energy 9×10^{-3} J. The displacement of the particle from its equilibrium position is 60 cm at time $\pi/4$ s. Calculate the amplitude of oscillations and mass of the particle.

Solution: Time period $T = 2\pi$ s. Total energy $E = 9 \times 10^{-3}$ J. Displacement $x = 60$ cm = 0.6 m at time $t = \pi/4$ s. For simple harmonic motion, the displacement is

$$x = A \sin(\omega t) = A \sin\left(\frac{2\pi}{T} t\right)$$

Thus, we have

$$0.6 = A \sin\left(\frac{2\pi}{2\pi} \frac{\pi}{4}\right) = \frac{A}{\sqrt{2}} \qquad \text{or} \qquad A = 0.6\sqrt{2}$$

The amplitude of oscillations is $A = 0.6\sqrt{2} = 0.85$ m. Mass of the particle is

$$m = \frac{ET^2}{2\pi^2 A^2} = \frac{9 \times 10^{-3} \times 4\pi^2}{2\pi^2 \times 0.6^2 \times 2} = 2.5 \times 10^{-2}\ \text{kg} = 25\ \text{g}$$

Exercise 5.5: A particle is executing the simple harmonic motion with a time period of 30 s and amplitude 15 cm. Calculate the angular frequency of motion and the value of maximum velocity of the particle.

Solution: We have time period $T = 30$ s and amplitude $A = 15$ cm = 0.15 m. Angular frequency of the particle is

$$\omega = \frac{2\pi}{T} = \frac{2 \times 3.14}{30} = 0.209 \text{ rad/s}$$

Maximum velocity v_m of the particle is

$$v_m = A\omega = 0.15 \times \frac{2 \times 3.14}{30} = 0.0314 \text{ m/s}$$

Exercise 5.6: A particle of mass 1 kg, attached to a massless spring is allowed to execute simple harmonic motion with an amplitude of 2 m. During the motion, when the particle is at the maximum displacement, its total energy is 8 J. Calculate the force constant of the spring and time period of motion.

Solution: Amplitude of oscillations $A = 2$ m. Total energy $E = 8$ J. We, therefore, have

$$E = \frac{1}{2}kA^2 \qquad \text{or} \qquad 8 = \frac{1}{2}k \times 2^2 \qquad \text{or} \qquad k = 4 \text{ J/m}$$

The time period

$$T = 2\pi\sqrt{\frac{m}{k}} = 2\pi\sqrt{\frac{1}{4}} = \pi = 3.14 \text{ s}$$

Exercise 5.7: A particle is executing simple harmonic motion of amplitude 80 cm and time period 6 s. Find out the time taken by the particle in moving from one end of its path to a position 40 cm away from the mean position on the same side.

Solution: We have amplitude $A = 80$ cm = 0.8 m, time period $T = 6$ s and displacement $x = 40$ cm = 0.4 m. The displacement of a particle executing simple harmonic motion is

$$x = A\sin(\omega t + \delta)$$

Let initially the particle be at one end so that at $t = 0$, we have $x = A$. Then, we have

$$A = A\sin(\delta) \qquad \text{or} \qquad \sin(\delta) = 1 = \sin\pi/2$$

Thus, we have $\delta = \pi/2$. Now, the displacement of a particle is

$$x = A\sin(\omega t + \pi/2) = A\cos\omega t = A\cos\left(\frac{2\pi}{T}t\right)$$

Using values of x and T here, we have

$$0.4 = 0.8\cos\left(\frac{2\pi}{6}t\right) \qquad \text{or} \qquad \cos\left(\frac{\pi}{3}t\right) = \frac{1}{2} = \cos\left(\frac{\pi}{3}\right)$$

Therefore, we have

$$\frac{\pi}{3}t = \frac{\pi}{3} \qquad \text{or} \qquad t = 1\text{ s}$$

Exercise 5.8: Find the maximum velocity and maximum acceleration of a particle executing simple harmonic motion of time period 4π s and amplitude 2 m.

Solution: We have amplitude $A = 2$ m and time period $T = 4\pi$ s. The angular frequency is

$$\omega = \frac{2\pi}{T} = \frac{2\pi}{4\pi} = 0.5\text{ rad/s}$$

The maximum velocity is

$$v_{\text{m}} = A\omega = 2 \times 0.5 = 1\text{ m/s}$$

The magnitude of maximum acceleration is

$$a_{\text{m}} = A\omega^2 = 2 \times 0.5^2 = 0.5\text{ m/s}^2$$

5.3 Simple pendulum

An ideal simple pendulum consists of a heavy point mass suspended by a massless, inextensible and perfectly flexible string from a rigid support. In practice, such requirements cannot be fulfilled. Thus, a simple pendulum consists of a small heavy metallic sphere (generally called the bob) suspended by a long fine string from a rigid support. The arrangement of simple pendulum is shown in Figure 5.1 where the equilibrium position of the pendulum is AO. When the bob is displaced from its equilibrium position and released, it oscillates to and fro about its equilibrium position. The motion of the mass is said to be the simple harmonic motion. Suppose at any instant t, it is at the point P, which is at a distance x from its equilibrium position, and the string makes a small angle α about the direction AO. The length of the string is l.

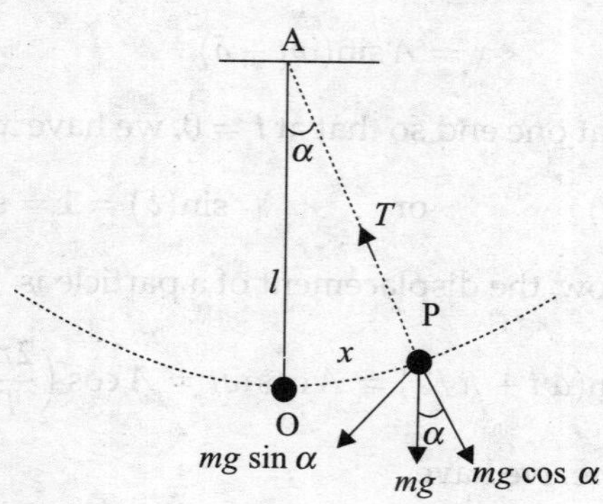

Figure 5.1: Simple pendulum.

At point P, the particle is under the action of two forces: (i) the weight mg of the bob acting vertically downward and (ii) the tension T in the string along PA. The weight mg can be resolved into two components: (i) the radial component $mg\cos\alpha$ which is working in the direction opposite to T and (ii) the transverse component $mg\sin\alpha$ working perpendicular to the direction PA. As the string is inextensible, we have $T = mg\cos\alpha$. The force $mg\sin\alpha$ tends to bring the bob to its mean (equilibrium) position. Hence, $mg\sin\alpha$ is the restoring force and we have

$$F = -mg\sin\alpha \tag{5.8}$$

The negative sign here indicates that the acceleration and displacement are oppositely directed. If x be the displacement at any time, the acceleration is $\mathrm{d}^2x/\mathrm{d}t^2$ and force is

$$F = m\,\frac{\mathrm{d}^2x}{\mathrm{d}t^2} \tag{5.9}$$

From equations (5.8) and (5.9), we have

$$m\,\frac{\mathrm{d}^2x}{\mathrm{d}t^2} = -mg\sin\alpha \qquad \text{or} \qquad \frac{\mathrm{d}^2x}{\mathrm{d}t^2} = -g\sin\alpha \tag{5.10}$$

For small angle α, we have[1]

$$\sin\alpha = \alpha = \frac{x}{l} \tag{5.11}$$

Using equation (5.11) in (5.10), we get

$$\frac{\mathrm{d}^2x}{\mathrm{d}t^2} + \frac{g}{l}\,x = 0 \qquad \text{or} \qquad \frac{\mathrm{d}^2x}{\mathrm{d}t^2} + \omega^2 x = 0$$

where $\omega = \sqrt{g/l}$. The time period T of oscillation is

$$T = \frac{2\pi}{\omega} = 2\pi\sqrt{\frac{l}{g}}$$

It is interesting to notice that the time period of the motion of the bob is independent of its mass and it depends only on the length of the string.

Exercise 5.9: A mass of 2 kg is attached to a spring of force constant 8 N/m. The system is allowed to make simple harmonic motion. Find the natural frequency of oscillations.

Solution: We have mass $m = 2$ kg and force constant $k = 8$ N/m. The natural frequency of oscillations is

$$n = \frac{1}{2\pi}\sqrt{\frac{k}{m}} = \frac{1}{2\pi}\sqrt{\frac{8}{2}} = \frac{1}{\pi}\ \text{Hz}$$

[1] We have

$$\sin\alpha = \alpha + \frac{\alpha^3}{3!} + \frac{\alpha^5}{5!} + \dots$$

For small value of α, we can neglect the terms having higher powers of α in comparison to α and thus, we have $\sin\alpha = \alpha$.

Exercise 5.10: A simple pendulum of 1.2 m length, hanging vertically at one end, is set to make simple harmonic motion. Calculate the time period of the oscillations of pendulum.

Solution: We have length of pendulum $l = 1.2$ m and acceleration due to gravity $g = 9.8$ m/s^2. The time period of oscillations is

$$T = 2\pi\sqrt{\frac{l}{g}} = 2 \times 3.14\sqrt{\frac{1.2}{9.8}} = 2.2 \text{ s}$$

Exercise 5.11: Time period of a simple pendulum is 4 s. Calculate the time period of the pendulum when its length is made four times that of its original length.

Solution: We have time period $T = 4$ s. Let the length of the pendulum be l. The we have

$$T = 2\pi\sqrt{\frac{l}{g}} \qquad \text{or} \qquad 4 = 2\pi\sqrt{\frac{l}{g}}$$

Let T' be the time period of oscillations when the length of the pendulum is increased to $4l$. Then, we have

$$T' = 2\pi\sqrt{\frac{4l}{g}}$$

From these equations, we have

$$\frac{T'}{4} = \frac{\sqrt{4l/g}}{\sqrt{l/g}} = 2$$

Thus, we have $T' = 8$ s.

Exercise 5.12: A simple pendulum of length 1 m is hanging in a lift which is ascending at an acceleration of 2.2 m/s^2. Calculate the period of oscillation of the pendulum.

Solution: We have length $l = 1$ m. The lift is ascending at an acceleration of 2.2 m/s^2. The acceleration due to gravity is 9.8 m/s^2. Thus, total acceleration is 9.8 + 2.2 = 12 m/s^2. The time period of oscillations is

$$T = 2\pi\sqrt{\frac{l}{g}} = 2\pi\sqrt{\frac{1}{12}} = 1.81 \text{ s}$$

5.4 Mass–string system

When a spring is compressed or stretched by applying a force on it, for small displacements, the elastic force (restoring force) is proportional to the displacement from the mean (equilibrium) position. Therefore,

$$F \propto -x \qquad \text{or} \qquad F = -kx$$

Here, the negative sign shows that the restoring force is always towards the equilibrium position and the proportionality constant k is known as the force constant or stiffness constant of spring. We can have two cases: (i) horizontal oscillations and (ii) vertical oscillations.

5.4.1 Horizontal oscillations

Let us consider a massless string lying in a horizontal plane and one end of which is connected to a point mass m and the other end is connected to a fixed point as shown in Figure 5.2. The mass m is free to move on a frictionless horizontal surface. When the mass is pulled to the right through a small distance x, then the restoring force exerted by the spring is directed towards the left and can be expressed as

$$F = -kx \tag{5.12}$$

Here, k is known as the force constant or stiffness constant of spring.

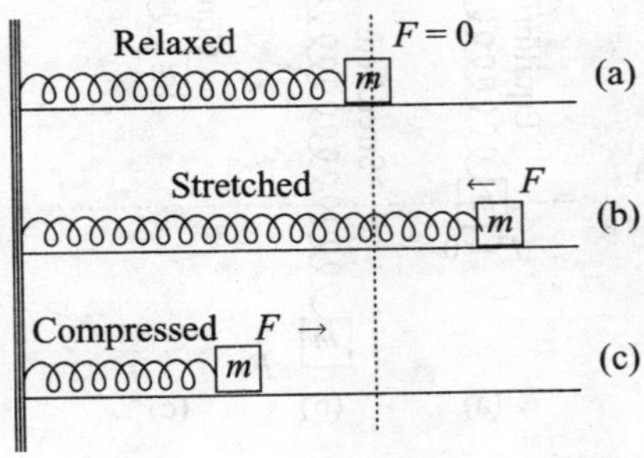

Figure 5.2: A body of mass m oscillates on a horizontal plane. In the position (a), no force acts on the spring; in position (b), the spring is stretched and the force acts in the left direction; in position (c), the spring is compressed and the force acts in the right direction.

Here, the negative sign shows that restoring force and displacement are oppositely directed. The mass moves with a linear acceleration $\mathrm{d}^2x/\mathrm{d}t^2$ and the force is

$$F = m\frac{\mathrm{d}^2x}{\mathrm{d}t^2} \tag{5.13}$$

From equations (5.12) and (5.13), we have

$$m\frac{\mathrm{d}^2x}{\mathrm{d}t^2} = -kx \qquad \text{or} \qquad \frac{\mathrm{d}^2x}{\mathrm{d}t^2} + \omega^2 x = 0$$

where $\omega = \sqrt{k/m}$ is generally known as the angular frequency. It is also known as the natural frequency of the oscillator. Thus, the time period T of oscillation is

$$T = \frac{2\pi}{\omega} = 2\pi\sqrt{\frac{m}{k}}$$

The time period of oscillations depend on the mass m and the force constant k.

5.4.2 Vertical oscillations

Let us consider a perfectly elastic and massless spring of length l hanging freely from a support (Figure 5.3). When a point mass m is attached to its lower end, the spring is stretched through some length due to gravitational force mg acting on the string in the downward direction. This situation corresponds to the equilibrium position of the system. Now, we pull down the mass through a small distance from the equilibrium position and then release. The mass starts oscillating to and fro about the equilibrium position. Suppose at any instant t, the displacement of the mass from the equilibrium position is x, the restoring force exerted by the spring in the upward direction can be expressed as

$$F = -kx \tag{5.14}$$

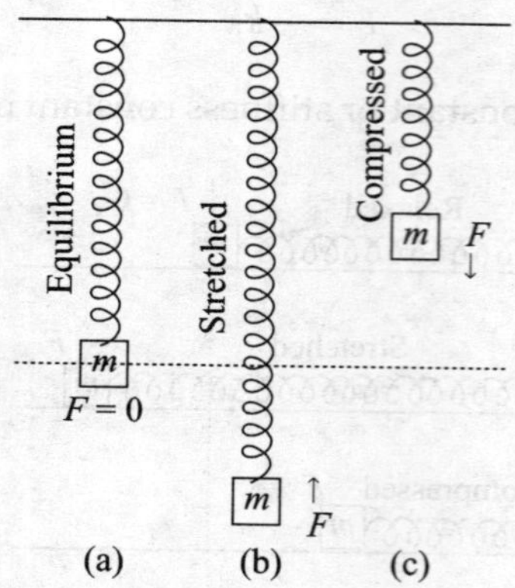

Figure 5.3: A body of mass m oscillates on a vertical plane. In the position (a), the spring is in the equilibrium position; in position (b), the spring is stretched and the force acts in the upward direction; in position (c), the spring is compressed and the force acts in the downward direction.

where k is the force constant or stiffness constant of spring. Here, the negative sign shows that restoring force and displacement are oppositely directed. The mass moves with a linear acceleration $\mathrm{d}^2x/\mathrm{d}t^2$ and the force is

$$F = m\,\frac{\mathrm{d}^2x}{\mathrm{d}t^2} \tag{5.15}$$

Thus, from equations (5.14) and (5.15), we have

$$m\,\frac{\mathrm{d}^2x}{\mathrm{d}t^2} = -kx \qquad \text{or} \qquad \frac{\mathrm{d}^2x}{\mathrm{d}t^2} + \omega^2 x = 0$$

where $\omega = \sqrt{k/m}$ is generally known as the angular frequency. It is also known as the natural frequency of the oscillator. Thus, the time period T of oscillations is

$$T = \frac{2\pi}{\omega} = 2\pi\sqrt{\frac{m}{k}}$$

The time period of oscillations depend on the mass m and the force constant k.

Exercise 5.13: A particle of mass 1 kg placed in a potential field $U = 32x^2 + 5$ J/kg is set to execute simple harmonic motion. Calculate the natural frequency of oscillations.

Solution: The potential field is $U = 32x^2 + 5$ and mass $m = 1$ kg. The force due to the potential field acting on the particle is

$$F = -\frac{\mathrm{d}U}{\mathrm{d}x} = -64x$$

For a particle of mass m, we have

$$m\frac{d^2x}{dt^2} = -64x \qquad \text{or} \qquad \frac{d^2x}{dt^2} = -\omega^2 x$$

where $\omega^2 = 64/m$. The natural frequency of oscillations is

$$\nu = \frac{\omega}{2\pi} = \frac{1}{2\pi}\sqrt{\frac{64}{m}} = \frac{1}{2\times 3.14}\sqrt{\frac{64}{1}} = 1.27\ \text{Hz}$$

Exercise 5.14: A body of mass 4 kg stretches a spring 0.2 m from its equilibrium position. The mass is replaced by another body of 2 kg and the system is set to make simple harmonic motion. Calculate the time period and frequency of oscillations.

Solution: For the body of mass 4 kg, we have $x = 0.2$ m. Thus, we have

$$F = kx \qquad \text{or} \qquad k = \frac{F}{x} = \frac{mg}{x} = \frac{4\times 9.8}{0.2} = 196\ \text{N/m}$$

For the body of mass 2 kg, the time period of oscillations is

$$T = 2\pi\sqrt{\frac{m}{k}} = 2\times 3.14\sqrt{\frac{2}{196}} = 0.63\ \text{s}$$

The frequency of oscillations is

$$\nu = \frac{1}{T} = \frac{1}{0.63} = 1.59\ \text{Hz}$$

Exercise 5.15: A body of mass 2 kg attached to a spring makes simple harmonic motion with time period 2.4 s. When the mass of the body is increased to 8 kg and set to make simple harmonic motion, calculate the time period of oscillations.

Solution: For the body of mass 2 kg, the time period is

$$T = 2\pi\sqrt{\frac{m}{k}} \qquad \text{or} \qquad 2.4 = 2\pi\sqrt{\frac{2}{k}}$$

For the body of mass 8 kg, the time period is

$$T' = 2\pi\sqrt{\frac{8}{k}}$$

From these equations, we have

$$\frac{T'}{2.4} = \frac{\sqrt{8/k}}{\sqrt{2/k}} = 2 \qquad \text{or} \qquad T' = 4.8\ \text{s}$$

5.5 Damped harmonic oscillator

In the preceding sections, we discussed about free oscillations where no frictional force or resistance is offered to the oscillating body. Consequently, the body keeps on oscillating indefinitely and such oscillations are known as the free oscillations. But, in real situations, there is always some resistance offered to the oscillating body. In real sense, when a body is set into oscillations, its amplitude continuously decreases due to fractional resistance working on the body and therefore the vibrations will die after some finite time. Such motion is said to be damped by the friction and is known as damped oscillations. Thus, when a pendulum is displaced from its equilibrium position, it oscillates with a decreasing amplitude and finally comes to rest.

Consider a body of mass m is attached to a spring whose force constant is k. Suppose at some time t, displacement of the body from its equilibrium position is x, the velocity of the body is $\mathrm{d}x/\mathrm{d}t$ and acceleration $\mathrm{d}^2x/\mathrm{d}t^2$. Here, two types of forces are acting on the body:

1. The restoring force $(-kx)$, which is proportional to the displacement x of the body from its mean position and is acting towards the mean position. Here, k is the force constant.

2. The damping force $(-q\,\mathrm{d}x/\mathrm{d}t)$, which is proportional to the velocity of the body and is directed opposite to the direction of the motion of the body. Here, q is coefficient of damping.

Hence, the total force F acting on the body is

$$F = -kx - q\,\frac{\mathrm{d}x}{\mathrm{d}t} \tag{5.16}$$

With acceleration $\mathrm{d}^2x/\mathrm{d}t^2$ of the body having mass m, the force is

$$F = m\,\frac{\mathrm{d}^2x}{\mathrm{d}t^2} \tag{5.17}$$

From equations (5.16) and (5.17), we have

$$m\,\frac{\mathrm{d}^2x}{\mathrm{d}t^2} = -kx - q\,\frac{\mathrm{d}x}{\mathrm{d}t}$$

or

$$\frac{\mathrm{d}^2x}{\mathrm{d}t^2} + 2s\,\frac{\mathrm{d}x}{\mathrm{d}t} + \omega^2 x = 0 \tag{5.18}$$

where $s = q/2m$ is the damping factor and $\omega = \sqrt{k/m}$ the angular frequency. This is differential equation of second order. Let us assume its solution to be of the form

$$x = A\,\mathrm{e}^{\alpha t} \tag{5.19}$$

so that

$$\frac{\mathrm{d}x}{\mathrm{d}t} = \alpha A\,\mathrm{e}^{\alpha t} \qquad \text{and} \qquad \frac{\mathrm{d}^2x}{\mathrm{d}t^2} = \alpha^2 A\,\mathrm{e}^{\alpha t} \tag{5.20}$$

Using equations (5.19) and (5.20) in (5.18), we have

$$\alpha^2 A\, e^{\alpha t} + 2s\alpha A\, e^{\alpha t} + \omega^2 A\, e^{\alpha t} = 0$$

or

$$A\, e^{\alpha t}(\alpha^2 + 2s\alpha + \omega^2) = 0$$

Since $A\, e^{\alpha t} \neq 0$, we have

$$\alpha^2 + 2s\alpha + \omega^2 = 0$$

This is a quadratic equation in α and its roots are

$$\alpha_1 = -s + \sqrt{s^2 - \omega^2} \quad \text{and} \quad \alpha_2 = -s - \sqrt{s^2 - \omega^2}$$

Therefore, the general solution of equation (5.18) is

$$x = A_1\, e^{(-s+\sqrt{s^2-\omega^2})t} + A_2\, e^{(-s-\sqrt{s^2-\omega^2})t} \tag{5.21}$$

where A_1 and A_2 are arbitrary constants. The actual solution depends on the situation whether (1) $s^2 > \omega^2$, (2) $s^2 = \omega^2$ or (3) $s^2 < \omega^2$.

Case 1: $s^2 > \omega^2$

Here, $\sqrt{(s^2 - \omega^2)}$ is real and less than s. Therefore, both the exponents $(-s + \sqrt{s^2 - \omega^2})$ and $(-s - \sqrt{s^2 - \omega^2})$ are negative. Further, $(-s - \sqrt{s^2 - \omega^2})$ is more negative as compared to $(-s + \sqrt{s^2 - \omega^2})$. Consequently, the factor $(-s - \sqrt{s^2 - \omega^2})$ will dominate and the displacement x decreases exponentially with time to zero without performing any oscillations. Such motion is said to be over-damped and may be expressed as

$$x = A_2\, e^{(-s-\sqrt{s^2-\omega^2})t}$$

Case 2: $s^2 = \omega^2$

Here, we have

$$x = (A_1 + A_2)\, e^{-st}$$

In this case also the displacement x decreases exponentially with time to zero without performing any oscillations. Such motion is known as critically damped, as it is the limit of the damped motion without oscillation.

Case 3: $s^2 < \omega^2$

Here, $\sqrt{s^2 - \omega^2}$ is imaginary and can be written as

$$\sqrt{s^2 - \omega^2} = i\beta$$

where $\beta = \sqrt{\omega^2 - s^2}$ and $i = \sqrt{-1}$. Thus, equation (5.21) becomes

$$x = e^{-st}[A_1 e^{i\beta t} + A_2 e^{-i\beta t}]$$

$$= e^{-st}[A_1(\cos\beta t + i\sin\beta t) + A_2(\cos\beta t - i\sin\beta t)]$$

$$= e^{-st}[(A_1 + A_2)\cos\beta t + i(A_1 - A_2)\sin\beta t] \tag{5.22}$$

Suppose $(A_1 + A_2) = A\sin\delta$ and $i(A_1 - A_2) = A\cos\delta$. Here, A and δ are constants. Then equation (5.22) becomes

$$x = e^{-st}[A\sin\delta\ \cos\beta t + A\cos\delta\ \sin\beta t] = A\,e^{-st}\cos(\beta t + \delta)$$

Substituting the value of β here, we have

$$x = A\,e^{-st}\cos[\sqrt{(\omega^2 - s^2)}\,t + \delta)]$$

This is equation of oscillatory motion with amplitude $A\,e^{-st}$. Obviously, the amplitude decreases with time t exponentially. However, the decay of amplitude depends on the damping factor s. Such motion is known as under-damped. Examples of such motion are the motion of a pendulum in air, or motion of the coil of the ballistic galvanometer, etc.

5.5.1 Energy decay

By definition, the energy of an oscillator is proportional to the square of the amplitude of motion. In the damped oscillator, the amplitude decays exponentially with time as e^{-st}. Hence, the energy of the oscillator decays as e^{-2st}. The decay of energy depends on the damping factor s and is fast for large value of s.

5.5.2 Logarithmic decrement

The rate at which the amplitude in the damped motion decays is measured through logarithmic decrement. The amplitude of damped harmonic motion is $A\,e^{-st}$. Therefore, at $t = 0$, the amplitude is maximum, $A = A_0$. Suppose $A_1, A_2, A_3, \ldots$ be the amplitudes at the time $t = T, 2T, 3T, \ldots$, respectively, where T is the time period of oscillations. Thus, we have

$$A_1 = A\,e^{-sT} \qquad A_2 = A\,e^{-2sT} \qquad A_3 = A\,e^{-3sT} \qquad \ldots$$

This gives

$$\frac{A_0}{A_1} = \frac{A_1}{A_2} = \frac{A_2}{A_3} = \ldots = e^{sT} = e^{\lambda}$$

Here, λ is known as logarithmic decrement. By taking natural logarithm, we have

$$\ln \frac{A_0}{A_1} = \ln \frac{A_1}{A_2} = \ln \frac{A_2}{A_3} = \ldots = \lambda$$

Thus, the logarithmic decrement is the natural logarithm of the ratio between two successive amplitudes which are separated by one time period.

5.5.3 Relaxation time

It is the time taken by a damped harmonic oscillator for decaying total mechanical energy by a factor 1/e of its initial value. The mechanical energy E of the oscillator with amplitude $A\,\mathrm{e}^{-st}$ is

$$E = \frac{1}{2} m\omega^2 (A\,\mathrm{e}^{-st})^2$$

At $t = 0$, we have

$$E_0 = \frac{1}{2} m\omega^2 A^2$$

Thus, we have

$$E = E_0\,\mathrm{e}^{-2st} \tag{5.23}$$

Suppose, τ be the relaxation time, then at $t = \tau$, we should have

$$E = \frac{E_0}{\mathrm{e}}$$

From equation (5.23), it gives

$$\frac{E_0}{\mathrm{e}} = E_0\,\mathrm{e}^{-2s\tau} \quad \text{or} \quad \mathrm{e} = \mathrm{e}^{2s\tau}$$

Therefore, we have

$$2s\tau = 1 \quad \text{or} \quad \tau = \frac{1}{2s}$$

Thus, equation (5.23) becomes

$$E = E_0\,\mathrm{e}^{-t/\tau}$$

This equation shows how the energy decays exponentially with time t.

Exercise 5.16: The relaxation time for a damped harmonic oscillator is 25 s. Calculate the time required to decrease the amplitude of the oscillator to 1/e times of initial value.

Solution: With initial amplitude A_0, the amplitude of damped harmonic oscillator is

$$A = A_0\,\mathrm{e}^{-st}$$

For the relaxation time τ, we have

$$\tau = \frac{1}{2s} \qquad \text{or} \qquad s = \frac{1}{2\tau} = \frac{1}{2 \times 25} = 0.02\ \text{s}^{-1}$$

Suppose after a time t', we have $A = A_0/\text{e}$. Then,

$$\frac{A_0}{\text{e}} = A_0\,\text{e}^{-st'} \qquad \text{or} \qquad \text{e} = \text{e}^{st'}$$

Thus,

$$st' = 1 \qquad \text{or} \qquad t' = \frac{1}{s} = \frac{1}{0.02} = 50\ \text{s}$$

5.6 Forced oscillations

In the preceding section, we have discussed oscillations where a body oscillates with its own (natural) frequency in absence of any external force. A different situation arises when an oscillating body is placed in an external periodic force. Such oscillations are known as the forced oscillations. In this case, the body finally oscillates under the influence of the external force and not with its natural frequency. So the forced oscillations can also be defined as the oscillations in which the body oscillates with frequency other than its natural frequency. This other frequency is due to some external periodic force. The tuning fork is one of the examples of the forced oscillations.

Suppose a body of mass m is connected to a massless, inextensible and perfectly flexible string. When it is displaced from its mean (equilibrium) position and released, it starts to oscillate to and fro about the mean position. Suppose at some time t, displacement of the body from its mean position is x, the velocity of the body is $\mathrm{d}x/\mathrm{d}t$ and acceleration $\mathrm{d}^2x/\mathrm{d}t^2$. Here, the body experiences three kinds of forces:

1. The restoring force $(-kx)$, which is proportional to the displacement x of the body from its mean position and is acting towards the mean position. Here, k is the force constant.

2. The damping force $(-q\,\mathrm{d}x/\mathrm{d}t)$, which is proportional to the velocity of the body and is directed opposite to the direction of the motion of the body. Here, q is coefficient of damping.

3. The external periodic force $F_0 \sin pt$ with angular frequency p.

Hence, the total force F acting on the body is

$$F = -kx - q\,\frac{\mathrm{d}x}{\mathrm{d}t} + F_0 \sin pt \tag{5.24}$$

With acceleration $\mathrm{d}^2x/\mathrm{d}t^2$ of the body, the force is

$$F = m\,\frac{\mathrm{d}^2x}{\mathrm{d}t^2} \tag{5.25}$$

From equations (5.24) and (5.25), we have

$$m\,\frac{d^2x}{dt^2} = -kx - q\,\frac{dx}{dt} + F_0 \sin pt$$

or

$$\frac{d^2x}{dt^2} + 2s\,\frac{dx}{dt} + \omega^2 x = f \sin pt \tag{5.26}$$

where $s = q/2m$ is the damping factor, $\omega = \sqrt{k/m}$ the angular frequency and $f = F_0/m$. Because of damping force, the body will lose its energy. The external force will however give energy periodically to the body, and finally, the body will oscillate with the frequency of the external force. Equation (5.26) is differential equation of second order. Since the body finally oscillates with the frequency of the applied external force. Let us assume that in the steady state, the solution of equation (5.26) be of the form

$$x = A \sin(pt - \delta) \tag{5.27}$$

where A is the amplitude of oscillations and δ is known as the phase angle. On differentiation of equation (5.27), we have

$$\frac{dx}{dt} = Ap\cos(pt-\delta) \qquad \text{and} \qquad \frac{d^2x}{dt^2} = -Ap^2 \sin(pt-\delta) \tag{5.28}$$

Using equations (5.27) and (5.28) in (5.26), we have

$$-Ap^2\sin(pt-\delta) + 2sAp\cos(pt-\delta) + \omega^2 A\sin(pt-\delta)$$

$$= f\sin[(pt-\delta)+\delta]$$

or

$$A\sin(pt-\delta)\,[\omega^2 - p^2] + 2sAp\cos(pt-\delta) = f\sin(pt-\delta)\ \cos\delta$$

$$+ f\cos(pt-\delta)\ \sin\delta$$

Comparing the coefficients of $\sin(pt-\delta)$ and $\cos(pt-\delta)$ on the two sides of this equation, we get

$$A(\omega^2 - p^2) = f\cos\delta \tag{5.29}$$

and

$$2sAp = f\sin\delta \tag{5.30}$$

Squaring equations (5.29) and (5.30) and adding them we get

$$A^2(\omega^2 - p^2)^2 + 4s^2A^2p^2 = f^2$$

or

$$A = \frac{f}{\sqrt{(\omega^2 - p^2)^2 + 4s^2p^2}} \tag{5.31}$$

Dividing equation (5.30) by (5.29), we get

$$\tan\delta = \frac{2sp}{(\omega^2 - p^2)}$$

or the phase angle

$$\delta = \tan^{-1}\left[\frac{2sp}{(\omega^2 - p^2)}\right] \tag{5.32}$$

Equations (5.31) and (5.32) show that both the amplitude and the phase angle of forced oscillations depend on $(\omega^2 - p^2)$. That is, they depend on both the driving frequency p and the natural frequency ω of the oscillator. The amplitude and phase angle for various situations are discussed in the following.

Case 1: Suppose the driving frequency p is very small as compared with the natural frequency ω, $p << \omega$. Then the amplitude of oscillations is

$$A = \frac{f}{\sqrt{(\omega^2 - p^2)^2 + 4s^2p^2}} \approx \frac{f}{\omega^2}$$

Using here the expressions for f and ω, we have

$$A = \frac{F_0/m}{k/m} = \frac{F_0}{k}$$

It shows that the amplitude of the forced oscillations depends on the force constant k of the spring and the magnitude F_0 of the applied external force. The phase angle is

$$\delta = \tan^{-1}\left[\frac{2sp}{(\omega^2 - p^2)}\right] = \tan^{-1}\left[\frac{2sp}{\omega^2}\right]$$

Since $\omega^2 >> p^2$, we have $2sp/\omega^2 \to 0$ and $\delta \to 0$. Thus, for this case, the driving force and the displacement are in phase.

Case 2: Suppose the driving frequency p is equal to the natural frequency ω, $p = \omega$. Such frequency is known as the resonant frequency. Then the amplitude of oscillations is

$$A = \frac{f}{\sqrt{(\omega^2 - p^2)^2 + 4s^2p^2}} = \frac{f}{2sp}$$

Substituting the expressions for f and s, we have

$$A = \frac{F_0/m}{(q/m)p} = \frac{F_0}{qp}$$

It shows that the amplitude A of forced oscillations depends on the damping coefficient q and the frequency p of applied force. The phase angle is

$$\delta = \tan^{-1}\left[\frac{2sp}{(\omega^2 - p^2)}\right] = \tan^{-1}[\infty] = \frac{\pi}{2}$$

Thus, the displacement lags behind the applied force by a phase angle of $\pi/2$ rad.

Case 3: Suppose the driving frequency is very large as compared with the natural frequency, $p >> \omega$. Then the amplitude of oscillations is

$$A = \frac{f}{\sqrt{(\omega^2 - p^2)^2 + 4s^2p^2}} = \frac{f}{\sqrt{p^4 + 4s^2p^2}} = \frac{f}{p^2}$$

Substituting the expression for f, we have

$$A = \frac{F_0}{mp^2}$$

It shows that the amplitude of the forced oscillations depends on the amplitude F_0 and frequency p of the applied force. The phase angle is

$$\delta = \tan^{-1}\left[\frac{2sp}{(\omega^2 - p^2)}\right] = \tan^{-1}\left[\frac{2sp}{-p^2}\right]$$

$$= \tan^{-1}\left[\frac{2s}{-p}\right] = \tan^{-1}[-0] = \pi$$

For this case, the displacement lags behind the force by a phase angle of π rad.

5.6.1 Resonance

In order to understand the phenomenon of resonance, we consider example of the forced oscillations which occur at the frequency of the external force with driving frequency p and not the natural frequency ω of the vibrating system. We have seen that the amplitude of oscillations depends on the relationship between p and ω. The amplitude is

$$A = \frac{f}{\sqrt{(\omega^2 - p^2)^2 + 4s^2p^2}}$$

For $p = 0$, we have

$$A_0 = \frac{f}{\omega^2}$$

and for $p = \omega$, we have

$$A_\omega = \frac{f}{2s\omega}$$

As we have

$$\frac{A_\omega}{A_0} = \frac{\omega}{2s}$$

It shows that ω must be larger than $2s$. For the given value of ω, when the damping factor s decreases, the amplitude A increases and the increase is the largest near $p = \omega$.

When there is no damping ($s = 0$), the amplitude becomes infinite at $p = \omega$. In Figure 5.4, we have shown variation of amplitude of forced oscillations as a function of p/ω for three values (small, medium and large) of damping factor s. The figure shows that the amplitude decreases when p is away from ω. The amplitude is maximum around $p = \omega$. The situation $p = \omega$ is known as the resonance. Further, when the damping is small, the amplitude of the forced oscillations increases rapidly as p approaches to ω. The amplitude reaches its maximum when $p = \omega$. For medium damping also, the amplitude increases but it does not increase so rapidly near $p = \omega$. As the damping increases, the resonance frequency is displaced slightly away from the natural frequency.

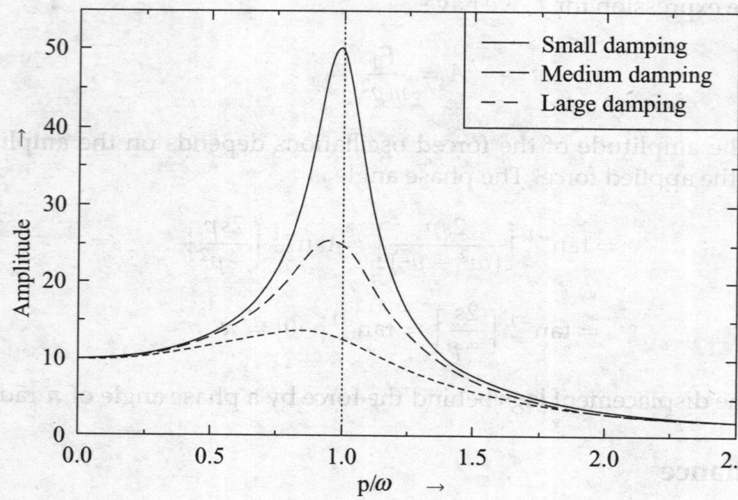

Figure 5.4: Variation of amplitude versus p/ω for three values of damping factor s.

Multiple-choice questions

1. For a periodic motion, the relation between time period T and frequency ν is

   (i) $T = 1/\nu$ (ii) $T = \nu^3$

   (iii) $T = \nu^2$ (iv) Gravitational force

   Ans. (i)

2. For a simple harmonic motion, which of the following quantities is conserved?

   (i) Kinetic energy (ii) Potential energy

   (iii) Total energy (iv) Angular momentum

   Ans. (iii)

3. For a simple pendulum, which of the following is wrong? Time period of a simple pendulum depends on

(i) mass of the bob. (ii) length of the massless thread.
(iii) the acceleration due to gravity. (iv) the person performing the experiment.

Ans. (i)

Exercises

1. Describe simple harmonic motion of a particle.
2. A particle of mass 200 g is executing simple harmonic motion with time period of π s. The displacement of the particle from its equilibrium position is 80 cm at time $\pi/8$ s. Calculate total energy of the particle.
3. A particle executes simple harmonic motion with time period of 20 s and amplitude 10 cm. Calculate maximum velocity of the particle.
4. A particle of mass m executes simple harmonic motion under a restoring force with force constant k. Obtain expressions for displacement, velocity and acceleration of the particle.
5. Suppose displacement x of a particle of mass m executing simple harmonic motion is

$$x = A\sin(\omega t + \delta)$$

Derive expressions for kinetic energy and potential energy of the particle. Show that the total energy of the particle is conserved.

6. Show that during simple harmonic motion, the kinetic energy of the particle changes into the potential energy of the particle and vice versa.
7. Derive expression for the time period of a simple pendulum.
8. Show that the time period of a simple pendulum does not depend on its mass.
9. Describe the horizontal motion of a mass attached to a perfectly elastic and massless spring.
10. Describe the vertical motion of a mass attached to a perfectly elastic and massless spring.
11. Show that for a simple harmonic oscillator, the average kinetic energy is equal to the average potential energy.
12. Describe the motion of a damped harmonic oscillator.
13. Describe the logarithmic decrement and relaxation time for a damped harmonic oscillator.
14. Describe the motion of an oscillator under an external force.

15. Show that for forced oscillations, when the frequency of the external force is much smaller than the natural frequency of oscillations, the displacement and the driving force are in phase.

16. Show that for forced oscillations, when the frequency of the external force is equal to the natural frequency of oscillations, the displacement lags behind the applied force by a phase angle of $\pi/2$ radians.

17. Show that for forced oscillations, when the frequency of the external force is much larger than the natural frequency of oscillations, the displacement lags behind the applied force by a phase angle of π rad.

18. Write notes on the following:

    (i) Harmonic motion
    (ii) Time period
    (iii) Frequency
    (iv) Simple pendulum
    (v) Damped harmonic oscillator
    (vi) Energy decay in a damped harmonic oscillator
    (vii) Logarithmic decrement in a damped harmonic oscillator
    (viii) Relaxation time in a damped harmonic oscillator
    (ix) Forced oscillations
    (x) Resonance in a forced oscillator

CHAPTER 6

THEORY OF RELATIVITY

Velocities of objects can be classified into two categories: (i) non-relativistic and (ii) relativistic. The velocities which are much smaller than the speed of light belong to the category of non-relativistic velocities and the velocities which are comparable to the speed of light belong to the category of relativistic velocities. A body moving with non-relativistic velocity is dealt under the classical mechanics. In the classical mechanics, energy and mass follow the conservation laws separately and there is no relation between the energy and mass. Further, the mass of a body does not depend on the velocity of the body. Moreover, the space and time are independent of each other. For transformations between two inertial reference frames, one has to follow the Galilean transformations. Suppose two bodies A and B are moving with velocities 10 km/hr in opposite directions. Then the velocity of A relative to that of B is 20 km/hr.

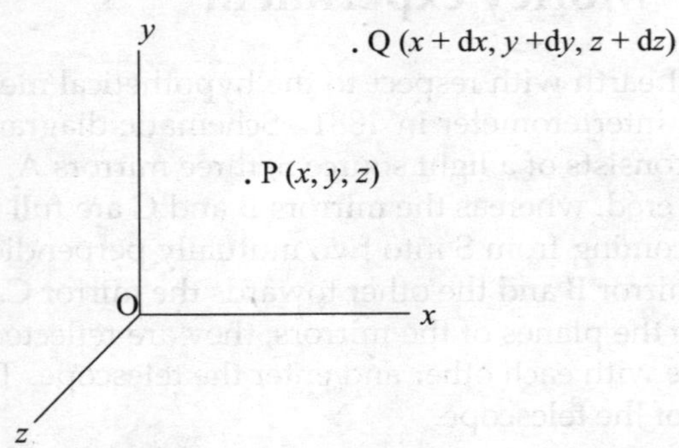

Suppose a body is at a point P whose coordinates are (x, y, z) at time t. The body moves with velocity v and at time $t + \mathrm{d}t$, it is at point Q whose coordinates are $(x + \mathrm{d}x,\ y + \mathrm{d}y,\ z + \mathrm{d}z)$. The separation PQ can be obtained as

$$\mathrm{PQ} = \sqrt{\mathrm{d}x^2 + \mathrm{d}y^2 + \mathrm{d}z^2} \tag{6.1}$$

as well as

$$\mathrm{PQ} = v\,\mathrm{d}t \tag{6.2}$$

Since the time and space are independent of each other, from equations (6.1) and (6.2), we have

$$v\,\mathrm{d}t = \sqrt{\mathrm{d}x^2 + \mathrm{d}y^2 + \mathrm{d}z^2} \quad \text{or} \quad v^2\mathrm{d}t^2 - \mathrm{d}x^2 - \mathrm{d}y^2 - \mathrm{d}z^2 = 0 \tag{6.3}$$

On the other side, when the velocity of a body is comparable to the speed of light, the motion is said to be relativistic. In the relativistic mechanics, energy and mass are no longer independent of each other. There is equivalence between energy and mass and we have conservation of energy and mass combined. The mass of the body varies with its velocity. The space and time are no longer independent of each other; rather the space and time are coupled with each other. The transformations between two inertial reference frames are governed by the Lorentz transformations. Now, suppose two bodies A and B are moving with velocities $0.8c$ (where c is the speed of light) in opposite directions. Then the velocity of A relative to that of B is not $1.6c$. Now, we have a four-dimensional space where coordinates are (ict, x, y, z). In the relativistic mechanics, the interval $\mathrm{d}s$ between the two points P (ict, x, y, z) and Q $(ic[t+\mathrm{d}t], x+\mathrm{d}x, y+\mathrm{d}y, z+\mathrm{d}z)$ is expressed as

$$\mathrm{d}s^2 = c^2\mathrm{d}t^2 - \mathrm{d}x^2 - \mathrm{d}y^2 - \mathrm{d}z^2 \tag{6.4}$$

Using equation (6.3) in (6.4), we get

$$\mathrm{d}s^2 = c^2\mathrm{d}t^2 - v^2\mathrm{d}t^2 \quad \text{or} \quad \mathrm{d}s = c\,\mathrm{d}t\sqrt{1-\beta^2} = \gamma c\,\mathrm{d}t$$

where $\beta = v/c$ and $\gamma = \sqrt{1-\beta^2}$.

6.1 Michelson - Morley experiment

To determine velocity of earth with respect to the hypothetical medium ether, Michelson and Morley devised an interferometer in 1881. Schematic diagram of the instrument is shown in Figure 6.1. It consists of a light source S, three mirrors A, B, C and a telescope T. The mirror A is half silvered, whereas the mirrors B and C are full silvered. The mirror A divides the light beam coming from S into two mutually perpendicular components; one travelling towards the mirror B and the other towards the mirror C. Since the directions of the beams are normal to the planes of the mirrors, they are reflected back. After retracing their paths they interfere with each other and enter the telescope. Thus, a fringe system is observed with the help of the telescope.

Suppose AB = l_1 and AC = l_2 and the velocity of the earth along AB is $\vec{v}$. The time taken by light to go from A to B and back from B to A is

$$t_1 = \frac{l_1}{c-v} + \frac{l_1}{c+v} = \frac{2l_1}{c(1-\beta^2)} \tag{6.5}$$

where $\beta = v/c$ and the time t_1 is measured from the earth's frame of reference. Let us now consider the beam moving in the direction of C. As seen from the ether's frame, the whole apparatus is moving with a velocity v in the direction of AB. Thus, as seen from the ether's frame, the light leaving A travels along the oblique path AC′, strikes the mirror at C′ and after reflection travels along C′A′ to reach the mirror at A′ (Figure 6.2). The distance travelled by light is now AC′ + C′A′. Let the time taken for this journey be t_2, as measured from the ether's frame. Now, AC′ = $c(t_2/2)$ and CC′ = $v(t_2/2)$. From the triangle ACC′, we have

$$l_2^2 + \left(\frac{vt_2}{2}\right)^2 = \left(\frac{c\,t_2}{2}\right)^2$$

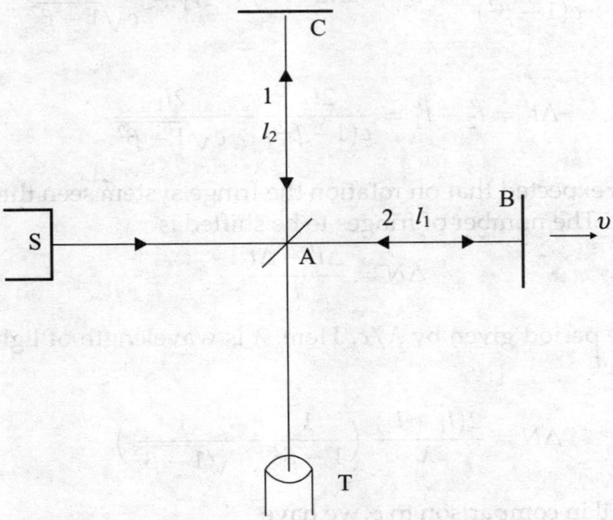

Figure 6.1: Schematic diagram of Michelson - Morley interferometer.

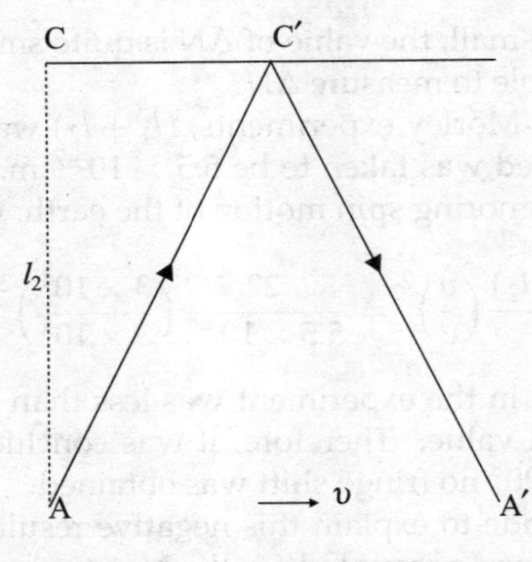

Figure 6.2: Perpendicular component as seen from the ether frame.

On solving, we get

$$t_2 = \frac{2l_2}{c\sqrt{1-\beta^2}} \tag{6.6}$$

We thus have

$$\Delta t = t_2 - t_1 = \frac{2l_2}{c\sqrt{1-\beta^2}} - \frac{2l_1}{c(1-\beta^2)}$$

Now, this time difference Δt will produce a path difference between the beams 1 and 2. This path difference produces a fringe system as seen through the telescope.

Now the whole apparatus is rotated through 90°, so that beam 1 travels at right angle to the motion of the earth. Now,

$$t_2' = \frac{2l_2}{c(1-\beta^2)} \quad \text{and} \quad t_1' = \frac{2l_1}{c\sqrt{1-\beta^2}}$$

Therefore,

$$\Delta t' = t_2' - t_1' = \frac{2l_2}{c(1-\beta^2)} - \frac{2l_1}{c\sqrt{1-\beta^2}}$$

Since $\Delta t \neq \Delta t'$, it is expected that on rotation the fringe system seen through the telescope would show a shift. The number of fringes to be shifted is

$$\Delta N = \frac{\Delta t' - \Delta t}{T}$$

where T is the time period given by λ/c. Here, λ is wavelength of light. Using values of $\Delta t, \Delta t'$ and T, we get

$$\Delta N = \frac{2(l_1+l_2)}{\lambda}\left(\frac{1}{1-\beta^2} - \frac{1}{\sqrt{1-\beta^2}}\right)$$

Since v is quite small in comparison to c, we have

$$\Delta N = \frac{2(l_1+l_2)}{\lambda}\left[\left(1+\beta^2\right) - \left(1+\beta^2/2\right)\right] = \frac{(l_1+l_2)}{\lambda}\beta^2$$

Since the value of β^2 is very small, the value of ΔN is quite small. However, with suitable values of l_1 and l_2, it is possible to measure ΔN.

In one of the Michelson - Morley experiments, $(l_1 + l_2)$ was taken to be 22 m and the wavelength λ of the light used was taken to be 5.5×10^{-7} m. The orbital velocity of the earth is about 3×10^4 m/s. Ignoring spin motion of the earth, we have

$$\Delta N = \frac{(l_1+l_2)}{\lambda}\left(\frac{v}{c}\right)^2 = \frac{22}{5.5\times10^{-7}}\left(\frac{3\times10^4}{3\times10^8}\right)^2 = 0.4$$

But the fringe shift observed in the experiment was less than 0.01 of a fringe, that is less than 1/40th of the predicted value. Therefore, it was concluded that on rotation of the experimental setup through 90° no fringe shift was obtained.

Several attempts were made to explain this negative result. However, all of them assumed the presence of ether and accounted for the Newtonian mechanics. None of these attempts could explain the negative result. In 1905, Einstein came with his bold postulates which gave birth to a new type of physics, now known as special theory of relativity.

6.2 Special theory of relativity

Einstein realized that it was of no use to make further attempts to patch up the existing theory. Therefore, he proposed an entirely new theory based on the following two postulates:

1. All physical laws are the same in all inertial (non-accelerated) reference frames.
2. The speed of light in the free space (vacuum) does not depend on the motion of the source and/or of the observer. That is, the speed of light is constant.

This constancy of the speed of light in all inertial reference frames makes the electromagnetic waves to behave differently from the mechanical waves which require a material medium for their propagation. The second postulate of Einstein can help in explaining the null result of the Michelson - Morley experiment. According to this postulate, the velocity of light to an observer on earth will also be c, as it is independent of the velocity, of the source as well as of the observer. Accordingly, whether the light travels in the direction AB or BA, its velocity has the value c.

Now, we have $t_1 = 2l_1/c$. For the perpendicular component, the path of light as seen by an observer on the earth would be ACA. Thus, we have $t_2 = 2l_2/c$. The difference in time is $\Delta t = t_2 - t_1 = 2(l_2 - l_1)/c$. It is now independent of v. This result can be obtained by taking $v = 0$ in equations (6.5) and (6.6). In the similar manner, for the rotation through 90°, we can obtain $\Delta t' = t_1' - t_2' = 2(l_1 - l_2)/c$. Therefore, $\Delta N = 0$. It shows that no fringe shift would be seen through the telescope.

6.2.1 Occurrence of two events

Another important consequence of the said postulates is regarding the concept of simultaneity. According to the Galilean transformations, when two events are simultaneous in one inertial reference frame, they are simultaneous in all the inertial reference frames. But according to the second postulate of Einstein, the simultaneity is only a relative concept. Two simultaneous events occurring in two different places for an observer in S will not be simultaneous for another observer in S′.

In order to illustrate this, let us consider two observers A and A′ in the inertial reference frames S and S′, respectively. Obviously the observer A′ is moving with a velocity $\vec{v}$ relative to A in the direction of x-axis. Suppose there are two flash lamps L_1 and L_2, equidistant from A (Figure 6.3). The observer A can fire the lamps L_1 and L_2 by some remote control device.

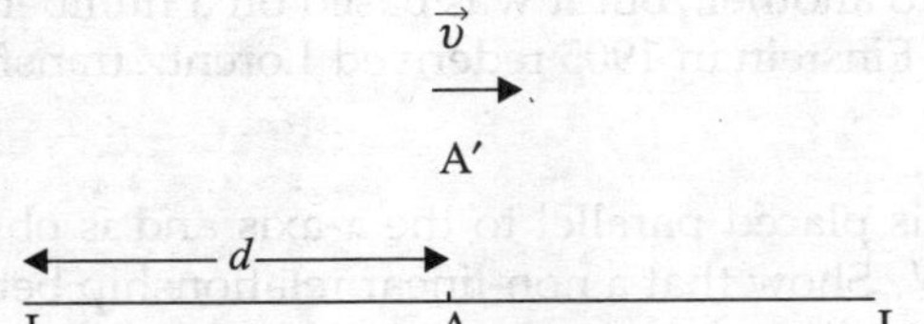

Figure 6.3: Relative nature of the simultaneity. The observer A′ is moving relative to the observer A with velocity $\vec{v}$ along the x-axis.

Suppose these lamps are fired simultaneously by A when A′ was overhead of A. Since the lamps are at equal distance from A, and A is not moving, the lights from both the lamps will reach to A at the same instance, so that the firing is simultaneous for him. Let us see what is the case with the observer A′. As the light takes finite time to reach A and during this time A′ would move towards L_2. The signal from L_2 would travel a shorter distance for reaching to A′ whereas the signal from L_1 would travel a longer distance for reaching to A′. Thus, A′ would see that the lamp L_2 was fired earlier than L_1. Hence, for the observer A′, the events are not simultaneous.

Suppose $L_1A = L_2A = d$. Then the time taken by light from L_1 and from L_2, to reach at A is $t = d/c$. During this time t, the observer A′ would move a distance $vt = vd/c$ towards lamp L_2. Thus, the time taken by light from L_2 to reach A′ is

$$t_2' = \frac{1}{c}\left(d - \frac{vd}{c}\right) = \frac{d}{c}\left(1 - \frac{v}{c}\right)$$

and the time t_1' taken by light from L_1 to reach to A′ is

$$t_1' = \frac{d}{c}\left(1 + \frac{v}{c}\right)$$

Here, we have assumed that the speed of light does not depend on the motion of the source and/or the observer (Einstein's second postulate). Thus, the time difference is

$$\Delta t' = t_1' - t_2' = \frac{2vd}{c^2}$$

It shows that the flashing of the two lamps L_1 and L_2 are simultaneous for the observer A whereas it is not simultaneous for the observer A′. This shows that the simultaneity is not absolute. That is, it is relative. Hence, the second postulate of Einstein is not consistent with the Galilean transformations, which tells that simultaneity is absolute.

The second postulate of Einstein intermixes the space and time coordinates and therefore it is no more correct to take $t = t'$, which has been true for Galilean transformations. Thus, a new transformation was needed which could accommodate the Einstein's second postulate. Further, under that transformation the electromagnetic laws should remain invariant in all the inertial reference frames. Consequently, the Galilean transformations are not the correct one.

For finding out such transformations, scientists could not succeed. Lorentz in 1904 derived a new transformation in which the Maxwell's equations were invariant from one inertial reference frame to another. But it was based on a number of ad hoc assumptions. Based on his postulates, Einstein in 1905 rederived Lorentz transformations, as discussed in the following section.

Exercise 6.1: A rod AB is placed parallel to the x-axis and is observed from the inertial reference frames S and S′. Show that a non-linear relationship between x and x' given by $x' = ax^2$ does not preserve the homogeneity of the space.

Solution: Let us consider a rod AB of length 3 m in the reference frame S, placed at $x_1 = 1$ m and $x_2 = 4$ m. When these coordinates are observed in S′, we have $x_1' = a(1)^2$ and $x_2' = a(4)^2$, so that the length of the rod observed in S′ is $x_2' - x_1' = 15a$ m.

Suppose the rod is shifted in S such that the new positions of A and B are at $x_1 = 2$ m and $x_2 = 5$ m. When these coordinates are observed in S′, we have $x_1' = a(2)^2$ and $x_2' = a(5)^2$, so that the length of the rod observed in S′ is $x_2' - x_1' = 21a$ m.

Thus, for a mere shifting of the rod in S, its length in S′ changes from $15a$ m to $21a$ m. This clearly shows that the space is in-homogeneous.

6.3 Lorentz transformations

Let us consider two inertial reference frames S and S′ moving with uniform velocities (Figure 6.4). Now, we consider a four-dimensional space in which a point P represents an event occurring at (x, y, z, t) when observed in S and at (x', y', z', t') when observed in S′. Here, the aim is to find out some transformations between the two coordinate systems, which satisfy Einstein's second postulate. Moreover, the transformation should be linear so that the homogeneity of the space is maintained.

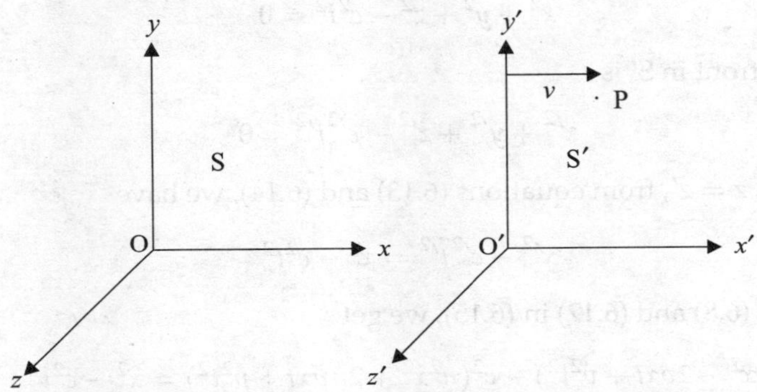

Figure 6.4: An inertial reference frame S′ is moving with velocity $\vec{v}$ relative to another inertial reference frame S along their common x-axis.

Since S′ is moving parallel to x-axis, we have $y = y'$ and $z = z'$. However, in the new transformation x and t are intermixed. Let us take x' and t' as linear combinations of x and t and write

$$x' = lx + mt \tag{6.7}$$

and

$$t' = nx + pt \tag{6.8}$$

Now, we need to find out the values of the coefficients l, m, n, p. Figure 6.4 shows that the origin O′ of reference frame S′ satisfies

$$x' = 0 \tag{6.9}$$

However, in the reference frame S, the motion of O′ is described as

$$x - vt = 0 \tag{6.10}$$

Since equations (6.9) and (6.10) both describe the motion of the same point O′, the two equations must have a linear relationship between them. Thus, we write

$$x' = q(x - vt) \tag{6.11}$$

Comparison of equations (6.7) and (6.11) shows that

$$l = q \qquad \text{and} \qquad m = -qv$$

Thus,

$$x' = l(x - vt) \tag{6.12}$$

In a similar manner, for the inverse transformation, we have

$$x = l(x' + vt')$$

Let a spherical electromagnetic wave is sent into the free space at $t = 0$ from the origins of S and S$'$. At some subsequent time, the equation of the wavefront in S is

$$x^2 + y^2 + z^2 - c^2t^2 = 0 \tag{6.13}$$

and of the wavefront in S$'$ is

$$x'^2 + y'^2 + z'^2 - c'^2t'^2 = 0 \tag{6.14}$$

Since $y = y'$ and $z = z'$, from equations (6.13) and (6.14), we have

$$x'^2 - c'^2t'^2 = x^2 - c^2t^2 \tag{6.15}$$

Using equations (6.8) and (6.12) in (6.15), we get

$$l^2(x^2 - 2vxt + v^2t^2) - c^2(n^2x^2 + 2npxt + p^2t^2) = x^2 - c^2t^2$$

Equating the coefficients of x^2, xt and t^2 on the two sides of this equation, we get

$$l^2 - c^2n^2 = 1 \tag{6.16}$$

$$vl^2 + c^2np = 0 \tag{6.17}$$

$$v^2l^2 - c^2p^2 = -c^2 \tag{6.18}$$

Solution of these equations (6.16), (6.17) and (6.18) is a bit tricky task. Equations (6.7) and (6.8) show that when there is no coupling between space and time, we have $m = n = 0$ and l may be equal to p, so that the transformation of space and of time has similar kind of variation. Using $l = p$ in equation (6.18), we get

$$l = p = \frac{1}{\sqrt{1-\beta^2}} \tag{6.19}$$

Using equation (6.19) in (6.16) or in (6.17), we get

$$n = -\frac{v}{c^2\sqrt{1-\beta^2}} \tag{6.20}$$

Using equation (6.19) in (6.12), we have

$$x' = \frac{x - vt}{\sqrt{1-\beta^2}}$$

Using equations (6.19) and (6.20) in (6.8), we get

$$t' = \frac{1}{\sqrt{1-\beta^2}}\left[t - \frac{\beta x}{c}\right]$$

Thus, finally, the Lorentz transformations from S to S′ are

$$x' = \frac{x - vt}{\sqrt{1-\beta^2}}; \qquad y' = y; \qquad z' = z; \qquad t' = \frac{t - \beta x/c}{\sqrt{1-\beta^2}} \tag{6.21}$$

The inverse Lorentz transformations from S′ to S are

$$x = \frac{x' + vt'}{\sqrt{1-\beta^2}}; \qquad y = y'; \qquad z = z'; \qquad t = \frac{t' + \beta x'/c}{\sqrt{1-\beta^2}} \tag{6.22}$$

It can be easily noted that when $v << c$, the Lorentz transformations reduce to the Galilean transformations and we have $t = t'$.

Exercise 6.2: Show that $x^2 + y^2 + z^2 - c^2t^2 = x'^2 + y'^2 + z'^2 - c^2t'^2$ under Lorentz transformations.

Solution: We can start from any set, equations (6.21) or (6.22), of Lorentz transformations. Using

$$x' = \frac{x - vt}{\sqrt{1-\beta^2}}; \qquad y' = y; \qquad z' = z; \qquad t' = \frac{t - \beta x/c}{\sqrt{1-\beta^2}}$$

we have

$$x'^2 + y'^2 + z'^2 - c^2t'^2 = \frac{(x-vt)^2}{1-\beta^2} + y^2 + z^2 - \frac{c^2(t-\beta x/c)^2}{1-\beta^2}$$

$$= \frac{(x^2 - 2xvt + v^2t^2) - (c^2t^2 - 2xvt + \beta^2x^2)}{1-\beta^2} + y^2 + z^2$$

$$= \frac{x^2(1-\beta^2) - c^2t^2(1 - v^2/c^2)}{1-\beta^2} + y^2 + z^2 = x^2 + y^2 + z^2 - c^2t^2$$

Exercise 6.3: An event occurs at the position (50, 20, 30) and time $t = 1 \times 10^{-4}$ s in an inertial reference frame S. Consider another inertial reference frame S′, which is moving with velocity $v = 2.4 \times 10^8$ m/s relative to that of S along the direction of x-axis. Calculate position and time of the event when measured in S′: (i) under Galilean transformations and (ii) under Lorentz transformations.

Solution: In S reference frame, we have

$$x = 50 \qquad y = 20 \qquad z = 30 \qquad t = 1 \times 10^{-4}$$

(i) The Galilean transformations are

$$x' = x - vt \qquad y' = y \qquad z' = z \qquad t' = t$$

Thus, we have

$$x' = 50 - 2.4 \times 10^8 \times 1 \times 10^{-4} = -23950$$

$$y' = 20 \qquad z' = 30 \qquad t' = 1 \times 10^{-4}$$

(ii) The Lorentz transformations are

$$x' = \frac{x - vt}{\sqrt{1-\beta^2}}; \qquad y' = y; \qquad z' = z; \qquad t' = \frac{t - \beta x/c}{\sqrt{1-\beta^2}}$$

We have $\beta = 2.4 \times 10^8/3 \times 10^8 = 0.8$. Using the values, we have

$$x' = \frac{x - vt}{\sqrt{1-\beta^2}} = \frac{50 - 2.4 \times 10^8 \times 1 \times 10^{-4}}{\sqrt{1-0.64}} = -9916.7$$

$$t' = \frac{t - \beta x/c}{\sqrt{1-\beta^2}} = \frac{1 \times 10^{-4} - 0.8 \times 50/3 \times 10^8}{\sqrt{1-0.64}} = 1.66 \times 10^{-4}$$

Thus, we have

$$x' = -9916.7 \qquad y' = 20 \qquad z' = 30 \qquad t' = 1.66 \times 10^{-4}$$

6.4 Length contraction

According to this, the length of an object moving with a velocity v relative to the observer appears to be contracted by a factor of $\sqrt{1 - v^2/c^2}$ along the direction of motion, where c is the speed of light in vacuum. The lengths in the directions perpendicular to the direction of motion remain unchanged.

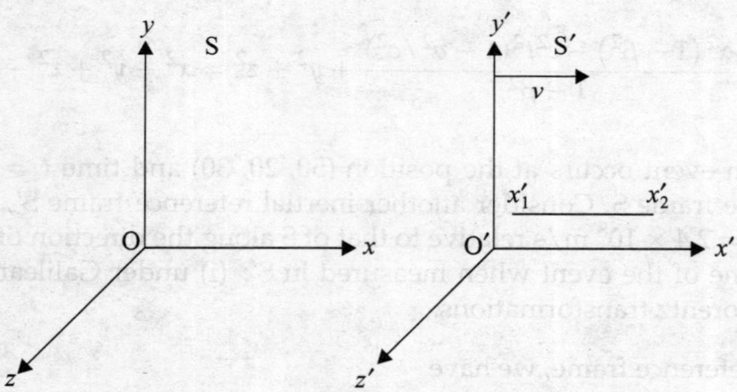

In order to derive expression for the length contraction, let us consider two inertial reference frames S and S′ where the latter is moving with uniform v relative to the first (Figure 6.4). Suppose a rigid rod of length l_0 is placed at rest in S′ along the x-axis. Thus, the velocity of rod relative to S′ is zero while it is moving with a velocity v relative to S in the positive direction of x-axis. Let x_1' and x_2' be the x-coordinates of the end points of the rod seen by an observer in S′. Therefore, we have

$$l_0 = x_2' - x_1'$$

Let x_1 and x_2 be the x-coordinates of the end points of the rod seen by an observer in S. Therefore, we have

$$l = x_2 - x_1$$

On applying the Lorentz transformations from S to S′ , we have

$$x_2' = \frac{x_2 - vt}{\sqrt{1 - v^2/c^2}} \quad \text{and} \quad x_1' = \frac{x_1 - vt}{\sqrt{1 - v^2/c^2}}$$

Thus, we have

$$x_2' - x_1' = \frac{x_2 - x_1}{\sqrt{1 - v^2/c^2}} \quad \text{or} \quad l_0 = \frac{l}{\sqrt{1 - v^2/c^2}}$$

Thus, $l < l_0$ shows that the length along the direction of motion of observer is contracted by a factor of $\sqrt{1 - v^2/c^2}$. Other Lorentz transformations are $y = y'$ and $z = z'$. It shows that the length of the rod in the direction perpendicular to the direction of motion is unchanged.

6.5 Time dilation

Another important significance of the special theory of relativity is the time dilation.

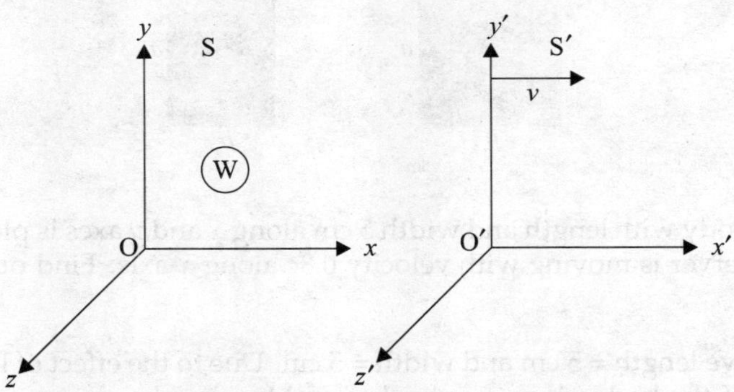

Suppose a clock W is situated at x in the inertial reference frame S and gives two signals at times t_1 and t_2. The time interval observed in S is $\Delta t = t_2 - t_1$. The times of the signals observed in S′ are

$$t_1' = \frac{t_1 - \beta x/c}{\sqrt{1 - \beta^2}} \quad \text{and} \quad t_2' = \frac{t_2 - \beta x/c}{\sqrt{1 - \beta^2}}$$

so that the time interval between the two signals observed in S′ is

$$\Delta t' = t_2' - t_1' = \frac{t_2 - t_1}{\sqrt{1 - \beta^2}} = \frac{\Delta t}{\sqrt{1 - \beta^2}}$$

Hence, the time interval in S′ is larger than that in S. The interval Δt appears to be dilated or lengthened, when it is observed by an observer in S′. Consequently, a clock will be found to run more and more slowly when the relative velocity between the clock and the observer increases more and more. The time interval between the two events measured by a clock at rest relative to an observer is known as the 'proper interval', whereas the time interval measured by an observer moving with a velocity relative to the clock is known as the improper interval.

Exercise 6.4: Show that under the Lorentz transformations, a square placed with sides along x and y axes in an inertial reference frame S will appear to be a rectangle when observed by an observer in the inertial reference frame S′ moving with a velocity V relative to S along the x-axis.

Solution: Let us consider two inertial reference frames S and S′ where the latter is moving with uniform velocity V relative to the first (see figure below). Suppose the sides of a square placed in S are parallel to x and y axes. Then the dimensions of the body in S parallel to x and y axes are equal to a, say. Due to the effect of length contraction, the dimension of the figure along x-axis observed by an observer in S′ will be

$$b = a\sqrt{1 - V^2/c^2}$$

The dimension along y-axis will remain unchanged as a. Therefore, the figure appears to be a rectangle with sides $a\sqrt{1 - V^2/c^2}$ and a as shown below.

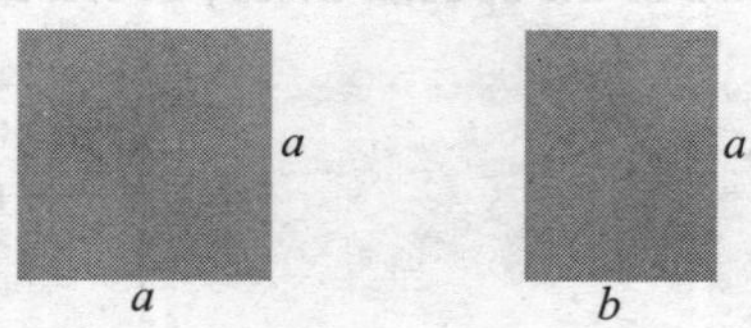

Exercise 6.5: A body with length and width 5 cm along x and y axes is placed in a reference frame S. An observer is moving with velocity 0.8c along x-axis. Find out the shape of the body.

Solution: We have length = 5 cm and width = 5 cm. Due to the effect of length contraction, the dimension of the body along x-axis observed by the observer moving with velocity $V = 0.8c$ will be

$$b = 5\sqrt{1 - (V/c)^2} = 5\sqrt{1 - (0.8c/c)^2} = 3 \text{ cm}$$

The dimension along y-axis will remain unchanged as 5 cm. Therefore, the figure appears to be a rectangle with sides of 5 cm and 3 cm.

Exercise 6.6: Show that under the Lorentz transformations, a circle placed in an inertial reference frame S appears to be an ellipse when observed by an observer in the inertial reference frame S′ moving with a velocity relative to S along x-axis.

Solution: Let us consider two inertial reference frames S and S′ where the latter is moving with uniform velocity V relative to the first (see figure). Suppose diameter of a circle placed in S is a. Due to the effect of length contraction, the dimension of the figure along x-axis observed by an observer in S′ will be

$$b = a\sqrt{1 - V^2/c^2}$$

The dimension along y-axis will remain unchanged as a. Therefore, the figure appears to be an ellipse with sides $a\sqrt{1 - V^2/c^2}$ and a as shown below.

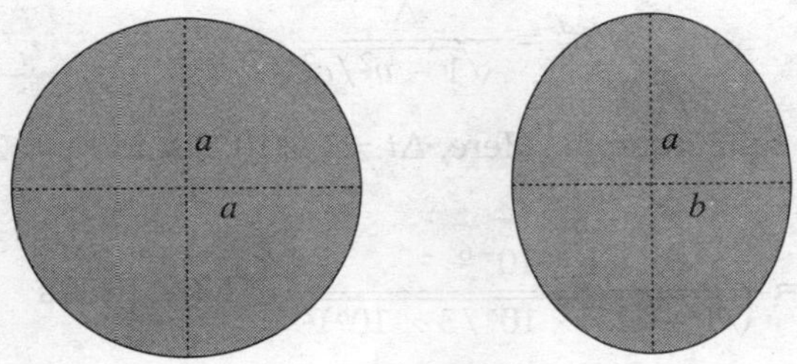

Exercise 6.7: A body having circular shape with radius 15 cm is placed in a reference frame S. An observer is moving with velocity $0.8c$ along x-axis. Find out the shape of the body.

Solution: We have diameter of circle 15 cm. Due to the effect of length contraction, the dimension of the body along x-axis observed by the observer moving with velocity $V = 0.8c$ will be

$$b = 15\sqrt{1 - (V/c)^2} = 15\sqrt{1 - (0.8c/c)^2} = 9 \text{ cm}$$

The dimension along y-axis will remain unchanged as 15 cm. Therefore, the figure appears to be an ellipse with axes 15 cm and 9 cm.

Exercise 6.8: The half-life of a radioactive particle measured in the laboratory is 4.0×10^{-8} s when its speed is $0.8c$. Calculate half-life of the particle when its speed is $0.6c$.

Solution: This phenomena of different half-lives for a particle moving with different velocities can be understood with the help of time dilation in the special theory of relativity. The time interval $\Delta t'$ measured by an observer in motion relative to the particle is

$$\Delta t' = \frac{\Delta t}{\sqrt{1 - (v/c)^2}}$$

where Δt is the proper time interval. In the first case, we have $\Delta t' = 4.0 \times 10^{-8}$ s and $v = 0.8c$. Thus, we have

$$\Delta t = 4.0 \times 10^{-8}\sqrt{1 - (0.8c/c)^2} = 2.4 \times 10^{-8} \text{ s}$$

Since the proper half-life is independent of velocity of the observer, the half-life of the particle when its speed is $0.6c$ is

$$\Delta t' = \frac{\Delta t}{\sqrt{1 - (v/c)^2}} = \frac{2.4 \times 10^{-8}}{\sqrt{1 - (0.6c/c)^2}} = 3 \times 10^{-8} \text{ s}$$

Exercise 6.9: The radioactive particle with a mean lifetime of 1 μs second is moving with a speed of 2.7×10^8 m/s relative to a laboratory.

1. What will be its lifetime as observed by an observer standing in the laboratory?
2. What will be the distance travelled by the particle before its disintegration?

Solution: The time interval $\Delta t'$ measured by an observer in motion relative to the particle is

$$\Delta t' = \frac{\Delta t}{\sqrt{1 - v^2/c^2}}$$

where Δt is the proper time interval. Here, $\Delta t = 1 \times 10^{-6}$ s and $v = 2.7 \times 10^8$ m/s. Thus, we have

$$\Delta t' = \frac{1 \times 10^{-6}}{\sqrt{1 - (2.7 \times 10^8 / 3 \times 10^8)^2}} = 2.3 \times 10^{-6}\ \text{s}$$

The distance travelled by the particle is

$$v\Delta t' = (2.7 \times 10^8) \times (2.3 \times 10^{-6}) = 621\ \text{m}$$

Exercise 6.10: At what speed a clock should be moved so that it may appear to lose 1 min in each hour.

Solution: When the clock is to lose 1 min in 1 hr, then the clock would record 59 min for each hour recorded by a clock which is stationary with respect to the observer. Suppose v be the required velocity of the clock. From the expression for time dilation, we have

$$\Delta t' = \frac{\Delta t}{\sqrt{1 - v^2/c^2}}$$

Using the values, we have

$$\frac{1}{59} = \frac{1/60}{\sqrt{1 - v^2/c^2}} \qquad \text{or} \qquad \frac{v}{c} = \sqrt{1 - (59/60)^2} = 0.182$$

Therefore, $v = 0.182 \times (3 \times 10^8) = 5.46 \times 10^7$ m/s.

Exercise 6.11: Show that the four-dimensional volume element dx dy dz dt remains invariant under Lorentz transformations.

Solution: In the Lorentz transformations, we have

$$x_1 = \frac{x_1' + vt_1'}{\sqrt{1 - \beta^2}} \qquad \text{and} \qquad x_2 = \frac{x_2' + vt_2'}{\sqrt{1 - \beta^2}}$$

Therefore,

$$x_2 - x_1 = \frac{x_2' - x_1'}{\sqrt{1 - \beta^2}} \qquad \mathrm{d}x = \frac{\mathrm{d}x'}{\sqrt{1 - \beta^2}}$$

This relation corresponds to the length contraction. Further,

$$y_2 - y_1 = y_2' - y_1' \qquad \text{or} \qquad \mathrm{d}y = \mathrm{d}y'$$

and

$$z_2 - z_1 = z'_2 - z'_1 \qquad \text{or} \qquad dz = dz'$$

In the inverse Lorentz transformations, we have

$$t'_1 = \frac{t_1 - vx/c^2}{\sqrt{1-\beta^2}} \qquad t'_2 = \frac{t_2 - vx/c^2}{\sqrt{1-\beta^2}}$$

Therefore,

$$t'_2 - t'_1 = \frac{t_2 - t_1}{\sqrt{1-\beta^2}} \qquad dt' = \frac{dt}{\sqrt{1-\beta^2}}$$

This relation corresponds to the time dilation. From the above relations, we have

$$dx\, dy\, dz\, dt = \frac{dx'}{\sqrt{1-\beta^2}} dy'\, dz'\, dt' \sqrt{1-\beta^2} = dx'\, dy'\, dz'\, dt'$$

It shows that the four-dimensional volume element is invariant under Lorentz transformations.

6.6 Lorentz transformations for velocity

Differentiating equation (6.21), we get

$$dx' = \frac{dx - vdt}{\sqrt{1-\beta^2}}; \qquad dy' = dy; \qquad dz' = dz; \tag{6.23}$$

$$dt' = \frac{dt - \beta dx/c}{\sqrt{1-\beta^2}} \tag{6.24}$$

Dividing equation (6.23) by (6.24), we get

$$\frac{dx'}{dt'} = \frac{dx - vdt}{dt - vdx/c^2} = \frac{dx/dt - v}{1 - \beta(dx/dt)/c}$$

$$\frac{dy'}{dt'} = \frac{dy(\sqrt{1-\beta^2})}{dt - vdx/c^2} = \frac{dy/dt(\sqrt{1-\beta^2})}{1 - \beta(dx/dt)/c}$$

$$\frac{dz'}{dt'} = \frac{dz(\sqrt{1-\beta^2})}{dt - vdx/c^2} = \frac{dz/dt(\sqrt{1-\beta^2})}{1 - \beta(dx/dt)/c}$$

Thus, the Lorentz transformation for velocity are

$$u'_x = \frac{u_x - v}{1 - \beta u_x/c}; \qquad u'_y = \frac{u_y(\sqrt{1-\beta^2})}{1 - \beta u_x/c}; \qquad u'_z = \frac{u_z(\sqrt{1-\beta^2})}{1 - \beta u_x/c}$$

where u'_x, u'_y, u'_z are the components of velocity measured in the reference frame S′ and u_x, u_y, u_z are the components of velocity measured in the reference frame S.

In the similar manner, from equation (6.22), we can get the inverse Lorentz transformation for velocity as

$$u_x = \frac{u'_x + v}{1 + \beta u'_x / c}; \qquad u_y = \frac{u'_y(\sqrt{1-\beta^2})}{1 + \beta u'_x / c}; \qquad u_z = \frac{u'_z(\sqrt{1-\beta^2})}{1 + \beta u'_x / c}$$

Exercise 6.12: An electron is moving with a speed of $0.8c$ in a direction opposite to that of a moving photon. Calculate the velocity of photon relative to the electron.

Solution: Suppose the photon and electron are moving along the positive and negative directions of x-axis. Let the electron moving with a velocity $-0.8c$ be at rest in S. Then we can say that the laboratory, which is S′, is moving with a velocity $0.8c$ relative to S (electron). Thus, we have $v = 0.8c$ and $u'_x = c$. Thus, the velocity of photon relative to the electron is

$$u_x = \frac{u'_x + v}{1 + v u'_x / c^2} = \frac{c + 0.8c}{1 + 0.8c(c)/c^2} = c$$

Exercise 6.13: Two particles are moving towards each other with speed $0.6c$ as observed by an observer standing on the earth. What is their relative speed?

Solution: Suppose the two particles are moving towards each other along the positive and negative directions of x-axis. Let one particle moving with a velocity $-0.6c$ be at rest in S. Then we can say that the earth, which is S′, is moving with a velocity $0.6c$ relative to S (first particle). Thus, we have $v = 0.6c$ and $u'_x = 0.6c$. Thus, the velocity of the second particle relative to the first is

$$u_x = \frac{u'_x + v}{1 + v u'_x / c^2} = \frac{0.6c + 0.6c}{1 + 0.6c(0.6c)/c^2} = 0.882c$$

Exercise 6.14: A particle is moving with a velocity $\vec{u}' = 0.1c\,\hat{i} + 0.2c\,\hat{j} - 0.4c\,\hat{k}$ when observed in the reference frame S′. Calculate velocity of the particle observed in the reference frame S, when S′ moves with a velocity $0.6c$ relative to S along the positive direction of x-axis.

Solution: For $\vec{u}' = 0.1c\,\hat{i} + 0.2c\,\hat{j} - 0.4c\,\hat{k}$, we have $u'_x = 0.1c$, $u'_y = 0.2c$, $u'_z = -0.4c$. The reference frame S′ is moving with a velocity $0.6c$ relative to S along the positive direction of x-axis. Thus, $v = 0.6c$. Using the velocity transformations

$$u_x = \frac{u'_x + v}{1 + v u'_x / c^2}; \qquad u_y = \frac{u'_y(\sqrt{1-\beta^2})}{1 + v u'_x / c^2}; \qquad u_z = \frac{u'_z(\sqrt{1-\beta^2})}{1 + v u'_x / c^2}$$

we have

$$u_x = \frac{u'_x + v}{1 + v u'_x / c^2} = \frac{0.1c + 0.6c}{1 + 0.6c(0.1c)/c^2} = 0.66c$$

$$u_y = \frac{u'_y(\sqrt{1-\beta^2})}{1 + v u'_x / c^2} = \frac{0.2c\,[\sqrt{1-(0.6c/c)^2}]}{1 + 0.6c(0.1c)/c^2} = 0.15c$$

$$u_z = \frac{u'_z(\sqrt{1-\beta^2})}{1 + v u'_x / c^2} = \frac{-0.4c\,[\sqrt{1-(0.6c/c)^2}]}{1 + 0.6c(0.1c)/c^2} = -0.30c$$

Hence, the velocity $\vec{u}$ of the particle in the reference frame S is $\vec{u} = 0.66c\,\hat{i} + 0.15c\,\hat{j} - 0.30c\,\hat{k}$.

6.7 Variation of mass with velocity

Let us consider two inertial reference frames S and S′ moving with uniform velocities (Figure 6.5). For simplicity, we consider that S′ is moving with velocity v relative to S along the common x-axis.

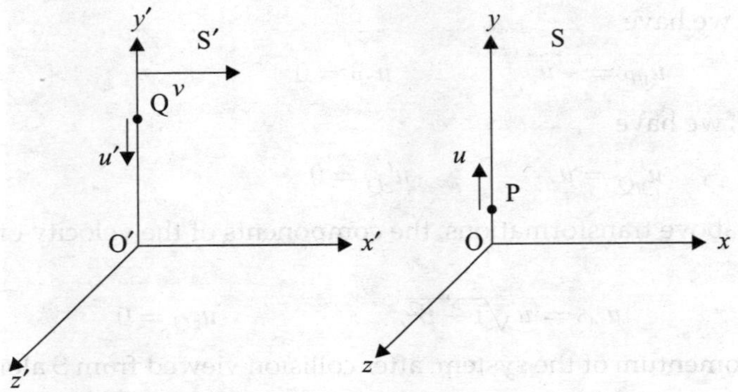

Figure 6.5: An inertial reference frame S′ is moving with velocity $\vec{v}$ relative to another inertial reference frame S along their common x-axis.

Let us consider two perfectly elastic balls P and Q of equal mass when measured in a common reference frame. The ball P is moving in S with velocity u along positive y-axis. The ball Q is moving in S′ with velocity u' along negative y-axis, such that $u' = -u$. Hence, the balls P and Q have equal and opposite velocities. Let the velocity u be non-relativistic, whereas the velocity v is relativistic. Let m_0 be the mass of each ball at rest. Suppose the balls collide, due to the motion of S′ relative to S, in such a way that the line of their centres is along the y-axis, so that the collision has no effect on the x and z components of velocity. As the collision is elastic, the directions of motion of each ball is reversed as seen in its own reference system. After collision, the velocity of P is $-u$ and that of Q is u, as in the elastic collision of two equal masses, the velocities are interchanged. Let us now analyse the collision process.

(i) Before collision

For the ball P, we have

$$u_{xP} = 0, \qquad u_{yP} = u, \qquad u_{zP} = 0$$

For the ball Q, we have

$$u'_{xQ} = 0, \qquad u'_{yQ} = -u, \qquad u'_{zQ} = 0$$

On applying transformations

$$v_{xQ} = \frac{v'_{xQ} + v}{1 + \beta v'_{xQ}/c}; \qquad v_{yQ} = \frac{v'_{yQ}(\sqrt{1-\beta^2})}{1 + \beta v'_{xQ}/c}; \qquad v_{zQ} = \frac{v'_{zQ}(\sqrt{1-\beta^2})}{1 + \beta v'_{xQ}/c}$$

the components of the velocity of Q measured in S are

$$u_{xQ} = v, \qquad u_{yQ} = -u\sqrt{1-\beta^2}, \qquad u_{zQ} = 0$$

Thus, the total momentum of the system, before collision viewed from S along y-axis is

$$m_0u - mu\sqrt{1-\beta^2}$$

where m is mass of the ball moving with velocity v.

(ii) After collision

For the ball P, we have

$$u_{xP} = 0, \qquad u_{yP} = -u, \qquad u_{zP} = 0$$

For the ball Q, we have

$$u'_{xQ} = 0, \qquad u'_{yQ} = u, \qquad u'_{zQ} = 0$$

On applying the above transformations, the components of the velocity of Q measured in S are

$$u_{xQ} = v, \qquad u_{yQ} = u\sqrt{1-\beta^2}, \qquad u_{zQ} = 0$$

Thus, the total momentum of the system, after collision viewed from S along y axis is

$$-m_0u + mu\sqrt{1-\beta^2}$$

Following the principle of the conservation of momentum, we have

$$m_0u - mu\sqrt{1-\beta^2} = -m_0u + mu\sqrt{1-\beta^2}$$

$$2m_0u = 2mu\sqrt{1-\beta^2}$$

$$m = \frac{m_0}{\sqrt{1-v^2/c^2}}$$

This relation shows that the mass m of a body increases with the increase of its velocity. It is obvious that no body having non-zero rest mass can move with a velocity equal to the speed of light.[1]

Exercise 6.15: Calculate the velocity at which the mass of a particle becomes double of its rest mass.

Solution: Mass of a moving body varies with its velocity according to the expression

$$m = \frac{m_0}{\sqrt{1-v^2/c^2}}$$

Here, we have to find out the value of v at which $m = 2m_0$.

$$2m_0 = \frac{m_0}{\sqrt{1-v^2/c^2}} \qquad 2 = \frac{1}{\sqrt{1-v^2/c^2}}$$

[1] A photon however moves with a velocity equal to c, because the rest mass of the photon is assumed to be zero. There is no contradiction of this assumption of zero rest mass of photon, as a photon is never at rest. Mass of a photon with frequency ν is $h\nu/c^2$ and the momentum of the photon is $h\nu/c$.

On solving, we have

$$v = \frac{\sqrt{3}}{2}c = \frac{1.732}{2} \times (3 \times 10^8) = 2.6 \times 10^8 \text{ m/s}$$

Thus, the mass of a particle moving with a velocity 2.6×10^8 m/s would become double of its rest mass.

Exercise 6.16: Rest mass of an electron is 9.1×10^{-31} kg. Calculate its mass when it is moving with a velocity $0.8c$.

Solution: Mass of the moving electron is

$$m = \frac{m_0}{\sqrt{1-(v/c)^2}} = \frac{9.1 \times 10^{-31}}{\sqrt{1-(0.8c/c)^2}} = 1.52 \times 10^{-30} \text{ kg}$$

6.8 Mass–energy equivalence

In relativistic mechanics, mass can be converted into energy and vice versa. Thus, mass and energy are equivalent to each other. A relation between mass and energy can be derived in the following manner. Suppose on application of an external force $\vec{F}$ on a body, having rest mass m_0, the velocity of the body increases from 0 to v. The amount of work done W is

$$W = \int_0^v \vec{F}\,.\,d\vec{s} = \int_0^v \vec{F}\,.\frac{d\vec{s}}{dt}\,dt = \int_0^v \frac{d\vec{p}}{dt}\,.\,\vec{v}\,dt$$

$$= \int_0^v \vec{v}\,.\frac{d(m\vec{v})}{dt}\,dt = \int_0^v \vec{v}\,.\,d(m\vec{v}) \tag{6.25}$$

This amount of work done appears in the form of kinetic energy, denoted by T. Mass m of a particle moving with velocity v is

$$m = \frac{m_0}{\sqrt{1-v^2/c^2}} \tag{6.26}$$

Using equation (6.26) in (6.25), we have

$$T = \int_0^v \vec{v}\,.\,d\left(\frac{m_0\vec{v}}{\sqrt{1-v^2/c^2}}\right)$$

$$= \int_0^v \left\{ \vec{v}\,.\frac{m_0}{(1-v^2/c^2)}\left[\sqrt{1-\frac{v^2}{c^2}}\,d\vec{v} - \frac{\vec{v}}{2}\left(1-\frac{v^2}{c^2}\right)^{-1/2}\left(-\frac{2\,\vec{v}\,.\,d\vec{v}}{c^2}\right)\right]\right\}$$

$$= \int_0^v \frac{m_0\vec{v}}{\left(1-v^2/c^2\right)^{3/2}}\,.\,d\vec{v}$$

This integral can be solved by using the substitution

$$y = 1 - \frac{v^2}{c^2} \quad \text{so that} \quad dy = -\frac{2\,\vec{v}}{c^2}.\,d\vec{v}$$

and we have

$$T = -\frac{m_0c^2}{2}\int_1^{1-v^2/c^2} \frac{dy}{y^{3/2}} = m_0c^2\left[\frac{1}{y^{1/2}}\right]_1^{1-v^2/c^2}$$

$$= m_0c^2\left[\frac{1}{\sqrt{1-v^2/c^2}} - 1\right] = mc^2 - m_0c^2$$

This is the relativistic expression for kinetic energy.[2] Using the energy–mass equivalence principle, the rest energy of the particle is m_0c^2. Hence, the total energy E of the particle is the sum of the relativistic kinetic energy and the rest energy. Thus, we have

$$E = T + m_0c^2 = mc^2 - m_0c^2 + m_0c^2 = mc^2$$

This is the well-known Einstein's energy–mass relation. It shows a universal equivalence between the mass and energy.

Conservation of energy and mass together

Owing to the equivalence between the mass and energy, the concept of consideration of mass and of energy separately is no more valid. Now, we have the conservation of energy and mass together.

Exercise 6.17: An electron with rest mass m_0 is moving with velocity $0.6c$, where c is the speed of light. Calculate the ratio of kinetic energy to rest energy of electron.

Solution: For rest mass m_0, the rest energy is

$$E_0 = m_0c^2$$

Kinetic energy of moving electron is

$$T = mc^2 - m_0c^2 = \frac{m_0c^2}{\sqrt{1-v^2/c^2}} - m_0c^2$$

[2] For a velocity $v << c$, the limiting value of kinetic energy is

$$T = mc^2 - m_0c^2 = \frac{m_0c^2}{\sqrt{1-v^2/c^2}} - m_0c^2 = m_0c^2(1 - v^2/c^2)^{-1/2} - m_0c^2$$

On expanding with the help of the Binomial theorem, we have

$$T = m_0c^2\left(1 + \frac{1}{2}\frac{v^2}{c^2} + \frac{3}{8}\frac{v^4}{c^4} + \ldots\right) - m_0c^2$$

Since $v << c$, we have

$$T = \frac{1}{2}\,m_0v^2$$

This is a well-known expression for kinetic energy of a non-relativistic particle. Thus, for a non-relativistic case, the kinetic energy is $\frac{1}{2}mv^2$ (Here, mass does not vary with velocity) and for a non-relativistic case, the kinetic energy is $(m - m_0)c^2$.

Using $v = 0.6c$, we get

$$T = \frac{m_0c^2}{\sqrt{1-(0.6c/c)^2}} - m_0c^2 = 1.25m_0c^2 - m_0c^2 = 0.25m_0c^2$$

The ratio of kinetic energy to rest energy is

$$\frac{T}{E_0} = \frac{0.25m_0c^2}{m_0c^2} = 0.25$$

Exercise 6.18: Calculate the amount of energy released on the conversion of 1 amu mass. (Given 1 amu = 1.66054×10^{-27} kg; $c = 3 \times 10^8$ m/s.)

Solution: The amount of energy equivalent to the mass is

$$E = mc^2 = 1.66054 \times 10^{-27} \times (3 \times 10^8)^2 = 1.4945 \times 10^{-10}\text{J}$$

$$= \frac{1.4945 \times 10^{-10}}{1.602 \times 10^{-19} \times 10^6} = 932.9 \text{ MeV}$$

Exercise 6.19: Calculate the rest energy of electron and proton in MeV. (Given rest mass of electron = 9.11×10^{-31} kg; rest mass of proton = 1.673×10^{-27} kg; $c = 3 \times 10^8$ m/s.)

Solution: The rest energy is the energy corresponding to the rest mass of the particle. For electron, we have

$$E = m_0c^2 = 9.11 \times 10^{-31} \times (3 \times 10^8)^2 = 8.199 \times 10^{-14}\text{J}$$

$$= \frac{8.199 \times 10^{-14}}{(1.602 \times 10^{-19}) \times 10^6} = 0.51 \text{ MeV}$$

For proton, we have

$$E = m_0c^2 = 1.673 \times 10^{-27} \times (3 \times 10^8)^2 = 1.5057 \times 10^{-10}\text{J}$$

$$= \frac{1.5057 \times 10^{-10}}{(1.602 \times 10^{-19}) \times 10^6} = 939.9 \text{ MeV}$$

Exercise 6.20: An electron and a positron at rest come together and annihilate, producing two photons of equal energy. Calculate the wavelength of each photon. (Given rest mass of electron as well of positron is 9.11×10^{-31} kg; $c = 3 \times 10^8$ m/s; $h = 6.626 \times 10^{-34}$ J s.)

Solution: Total mass of one electron and one positron at rest is

$$m = 2 \times (9.11 \times 10^{-31}) = 1.822 \times 10^{-30} \text{ kg}$$

Owing to annihilation of one electron and one positron at rest, the amount of energy produced is

$$E = mc^2 = 1.822 \times 10^{-30} \times (3 \times 10^8)^2 = 1.64 \times 10^{-13}\ \text{J}$$

This energy is equally divided into two photons. Thus, the energy of each photon is 8.2×10^{-14} J. The corresponding wavelength of each photon is

$$\lambda = \frac{hc}{E} = \frac{6.626 \times 10^{-34} \times (3 \times 10^8)}{8.2 \times 10^{-14}} = 2.42 \times 10^{-12}\ \text{m} = 0.0242\ \text{Å}$$

6.8.1 Momentum of a relativistic particle

When a particle moves with a velocity v comparable to the speed of light c, the motion is said to be relativistic. For the relativistic motion, mass of the particle varies with the velocity. Momentum of a particle moving with velocity v is

$$p = mv = \frac{m_0 v}{\sqrt{1 - v^2/c^2}}$$

where m_0 is the rest mass of the particle.

Exercise 6.21: Find the velocity of an electron so that its momentum is 10 times that of the product of its rest mass and the speed of light. Calculate the energy of the electron at this speed.

Solution: We are given $p = 10m_0c$. Thus, we have

$$p = \frac{m_0 v}{\sqrt{1 - v^2/c^2}} \qquad 10m_0 c = \frac{m_0 v}{\sqrt{1 - v^2/c^2}}$$

$$\frac{v}{c} = 10\sqrt{1 - \left(\frac{v}{c}\right)^2} \qquad \frac{v^2}{c^2} = 100\left[1 - \left(\frac{v}{c}\right)^2\right]$$

$$\frac{v^2}{c^2} = \frac{100}{101} \qquad \frac{v}{c} = \sqrt{\frac{100}{101}} = 0.995$$

Thus, velocity of the particle is $v = 0.995c = 0.995 \times (3 \times 10^8) = 2.985 \times 10^8$ m/s. The mass of the moving electron is

$$m = \frac{m_0}{\sqrt{1 - v^2/c^2}} = \frac{9.11 \times 10^{-31}}{\sqrt{1 - 100/101}} = 9.16 \times 10^{-30}\text{kg}$$

Energy of the electron is

$$E = mc^2 = 9.16 \times 10^{-30} \times (3 \times 10^8)^2 = 8.244 \times 10^{-13}\ \text{J}.$$

Exercise 6.22: Calculate the kinetic energy of an electron moving with a velocity 0.98 times the speed of light. (Given rest mass of electron = 9.11×10^{-31} kg.)

Solution: Mass of the moving electron is

$$m = \frac{m_0}{\sqrt{1 - v^2/c^2}} = \frac{9.11 \times 10^{-31}}{\sqrt{1 - (0.98c/c)^2}} = 4.578 \times 10^{-30} \text{ kg}$$

The kinetic energy is

$$T = (m - m_0)c^2$$

$$= (4.578 \times 10^{-30} - 9.11 \times 10^{-31}) \times (3 \times 10^8)^2 = 3.3 \times 10^{-13}\text{J}$$

$$= \frac{3.3 \times 10^{-13}}{1.602 \times 10^{-19} \times 10^6} = 3.11 \text{ MeV}$$

Exercise 6.23: For a particle with rest mass m_0, derive the energy–momentum relation $E^2 - c^2p^2 = m_0^2c^4$, where E is the total energy and p is the momentum.

Solution: Suppose a particle having rest mass m_0 is moving with velocity v. Its mass is

$$m = \frac{m_0}{\sqrt{1 - v^2/c^2}}$$

Total energy E and momentum p of the particle are

$$E = mc^2 \qquad\qquad p = mv$$

Now, we have

$$E^2 - c^2p^2 = m^2c^4 - c^2m^2v^2 = m^2c^4\left(1 - \frac{v^2}{c^2}\right)$$

Using the expression for m, we have

$$E^2 - c^2p^2 = \frac{m_0^2}{1 - v^2/c^2}c^4\left(1 - \frac{v^2}{c^2}\right) = m_0^2c^4$$

6.8.2 Binding energy of a nucleus

One may at first assume that the mass of an atom should be equal to the sum of the masses of its constituent particles, that is electrons, protons and neutrons. But the mass of a stable atom was found to be less than the sum of the masses of its constituent particles. The decrease in the mass is known as 'mass defect'. This mass defect can be explained in the following manner. The nucleons in a stable nucleus are held together by strong attractive forces, and therefore some work need to be done in separating them from one another until they are at a very large distances from one another. In other words, the energy needs to be supplied to a nucleus in order to break it into its constituents, separated by very large distance from one another. The principle of equivalence of mass and energy confirms the idea that the amount of energy required to break up the atom into its constituents is equivalent to the mass defect. It can also be defined as the amount of mass which would be converted into energy if the particular atom is to be assembled out of its constituents. The energy equivalent of mass defect is thus a measure of the binding energy of the atom.

The constituents of the atom ${}_Z X^A$ are Z protons, Z electrons and $(A-Z)$ neutrons. The total mass of individual constituents is $Zm_e + Zm_p + (A-Z)m_n$, where m_e, m_p and m_n are, respectively, the masses of electron, proton and neutron. If $m(Z, A)$ is the mass of the atom ${}_Z X^A$, then the mass defect Δm is

$$\Delta m = [Zm_p + Zm_e + (A-Z)m_n] - m(Z, A)$$

As the binding energy B is the energy equivalent of the mass defect, we have

$$B = c^2 \Delta m = [Zm_p + Zm_e + (A-Z)m_n - m(Z, A)]c^2$$

Exercise 6.24: Calculate the binding energy of a deuteron ${}_1H^2$. The masses of a deuteron, a proton and a neutron in amu are 2.014102, 1.007825 and 1.008665, respectively. Given 1 amu = 931 MeV.

Solution: Mass of one proton and one neutron = 1.007825 + 1.008665 = 2.01649 amu

Mass defect Δm = 2.01649 − 2.014102 = 0.002388 amu

Binding energy of deuteron = 0.002388 × 931 = 2.223 MeV

Exercise 6.25: For the nuclear reaction

$$4\,{}_1H^1 \rightarrow {}_2He^4 + 2\beta^+ + 2\nu + Q$$

the masses of ${}_1H^1$, ${}_2He^4$ and β^+ are, respectively, 1.007825, 4.002603 and 0.000549 amu. Considering the neutrinos ν to be massless, calculate the amount of energy released per reaction.

Solution: Initial mass = 4 × 1.007825 = 4.0313 amu

Final mass = 4.002603 + 2 × 0.000549 = 4.003701 amu

Thus, the mass loss during the fusion process = 4.0313 − 4.003701 = 0.027599 amu

Hence, the energy released per fusion = 0.027599 × 931.3 = 25.70 MeV

6.8.3 Photon

Let us consider a photon of frequency ν. Its energy is $h\nu$. From the principle of energy−mass equivalence, the mass of photon is

$$mc^2 = h\nu \qquad \text{or} \qquad m = \frac{h\nu}{c^2}$$

The momentum of photon is

$$p = mc = \frac{h\nu}{c}$$

The mass and momentum of a photon obviously depend on its frequency. The mass of a γ-ray photon will be much larger as compared to that of a visible photon. We can safely assume that rest mass of a photon is zero, because as soon as a photon comes to rest, it no more exists and transfers its energy to some object. Look at the expression

$$m = \frac{m_0}{\sqrt{1 - v^2/c^2}}$$

For $v = c$, the mass m becomes infinite when m_0 is not zero. Thus, we have to take $m_0 = 0$ in case of a photon so that m is indeterminate.

Exercise 6.26: Calculate mass and momentum of a photon having wavelength of 5 Å. Given $h = 6.62 \times 10^{-34}$ J s; $c = 3 \times 10^8$ m/s.

Solution: Mass of photon is

$$m = \frac{h\nu}{c^2} = \frac{h}{c\lambda} = \frac{6.62 \times 10^{-34}}{3 \times 10^8 \times 5 \times 10^{-10}} = 4.41 \times 10^{-33}\ \text{kg}$$

The momentum of photon is

$$p = \frac{h\nu}{c} = \frac{h}{\lambda} = \frac{6.62 \times 10^{-34}}{5 \times 10^{-10}} = 1.32 \times 10^{-24}\ \text{kg-m/s}$$

6.8.4 Compton effect

The spectrum of x-rays scattered from a graphite block was found to have intensity peaks at two wavelengths in each direction, one at the same wavelength λ as the incident radiations and the other at a longer wavelength λ'. The wavelength λ' was found to depend on the direction of scattering, whereas the wavelength λ was independent of the direction of scattering. This effect is known as the Compton effect. It can be explained in the following manner.

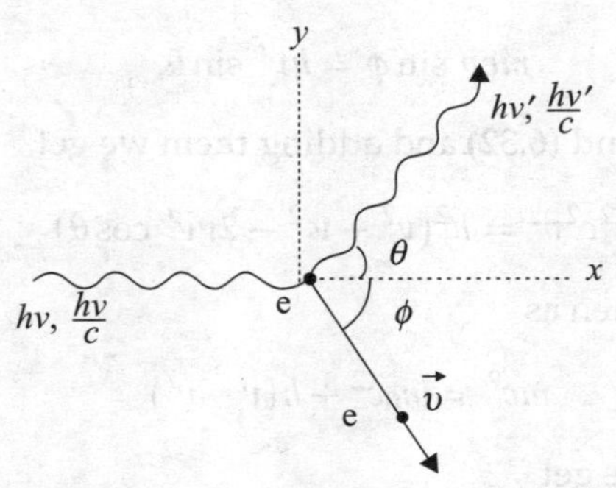

Figure 6.6: Photon of frequency ν moving along the x-axis incident on a free electron at rest placed at the origin of the coordinate system. After collision, the frequency of photon is ν' when scattered in a direction θ with the x-axis and the electron recoils with velocity $\vec{v}$ in a direction ϕ with the x-axis.

Let a photon of frequency ν incident on a free electron[3] at rest. For the photon having frequency ν, the energy is $h\nu$ and momentum is $h\nu/c$. For convenience, let the electron be

[3]Since the ionization potential of graphite is much smaller as compared to the energy of x-ray, an electron in the graphite may be considered as a free electron.

placed at the origin of the coordinate system and the photon is moving along the x-axis (Figure 6.6). Suppose, after collision the frequency of photon is ν' when it is scattered in the direction making an angle θ with the x-axis and the electron recoils with velocity $\vec{v}$ in a direction making an angle ϕ with the x-axis. For the scattered photon having frequency ν', the energy is $h\nu'$ and momentum is $h\nu'/c$.

Before collision, total energy and momentum of electron at rest are m_0c^2 and zero, respectively. Here, m_0 is the rest mass of the electron and c is the speed of light. It is a case of the relativistic motion and therefore mass of the electron varies with its velocity according to the relation

$$m = \frac{m_0}{\sqrt{1 - v^2/c^2}} \tag{6.27}$$

Using the principle of the conservation of energy, we have

$$h\nu + m_0c^2 = h\nu' + mc^2 \tag{6.28}$$

From the principle of the conservation of momentum along the x-axis and perpendicular to it, we have

$$\frac{h\nu}{c} = \frac{h\nu'}{c} \cos\theta + mv\cos\phi \tag{6.29}$$

and

$$0 = \frac{h\nu'}{c} \sin\theta - mv\sin\phi \tag{6.30}$$

Equations (6.29) and (6.30) can be written as

$$mcv \cos\phi = h(\nu - \nu' \cos\theta) \tag{6.31}$$

and

$$mcv \sin\phi = h\nu' \sin\theta \tag{6.32}$$

Squaring equations (6.31) and (6.32) and adding them we get

$$m^2c^2v^2 = h^2(\nu^2 + \nu'^2 - 2\nu\nu' \cos\theta) \tag{6.33}$$

Equation (6.28) can be written as

$$mc^2 = m_0c^2 + h(\nu - \nu') \tag{6.34}$$

Squaring equation (6.34) we get

$$m^2c^4 = m_0^2c^4 + h^2(\nu^2 + \nu'^2 - 2\nu\nu') + 2hm_0c^2(\nu - \nu') \tag{6.35}$$

Subtracting equation (6.33) from (6.35), we get

$$m^2c^2(c^2 - v^2) = m_0^2c^4 + h^2(-2\nu\nu' + 2\nu\nu' \cos\theta) + 2hm_0c^2(\nu - \nu')$$

Using equation (6.27) in this equation, we get

$$2hm_0c^2(\nu - \nu') = 2h^2\nu\nu'(1 - \cos\theta)$$

or

$$\frac{1}{\nu'} - \frac{1}{\nu} = \frac{h}{m_0 c^2}(1 - \cos\theta)$$

Using $\nu = c/\lambda$ and $\nu' = c/\lambda'$, we have

$$\Delta\lambda \equiv \lambda' - \lambda = \frac{h}{m_0 c}(1 - \cos\theta) \tag{6.36}$$

The value of θ can vary from zero to π rad. At $\theta = 0$, we have $\Delta\lambda = h/m_0 c$. This value is known as Compton wavelength. At $\theta = \pi$, we have $\Delta\lambda = 2h/m_0 c$. Thus, the wavelength λ' of the scattered radiations is longer than the wavelength λ of the incident radiations, and it obviously depends on the direction of scattering.

When instead of one free electron, the metal block participates in the scattering of radiations, in equation (6.36), m_0 is replaced by the mass M of the block. The value of M is very large and we have $\Delta\lambda = 0$, that is $\lambda' = \lambda$. Hence, the wavelength of the scattered radiations is same as that of the incident one. This explains the two wavelengths in the intensity peaks of the scattered light.

Exercise 6.27: X-rays of wavelength $\lambda = 1$ Å are scattered by a free electron. The scattered radiations are observed at an angle 90° relative to the direction of incident photon. Calculate the Compton shift. Given $h = 6.62 \times 10^{-34}$ J s; $m_0 = 9.11 \times 10^{-31}$ kg; $c = 3 \times 10^8$ m/s.

Solution: The Compton shift in wavelength is

$$\Delta\lambda = \frac{h}{m_0 c}(1 - \cos\theta)$$

$$= \frac{6.62 \times 10^{-34}}{9.11 \times 10^{-31} \times 3 \times 10^8}(1 - \cos 90°)$$

$$= 2.42 \times 10^{-12} \text{ m} = 0.0242 \text{ Å}$$

Exercise 6.28: X-rays of wavelength $\lambda = 1$ Å are scattered by a carbon block. The scattered radiations are detected at a right angle to the direction of incident photon. Calculate (i) Compton shift in wavelength, and (ii) the energy imparted to the recoiling electron. Given $h = 6.62 \times 10^{-34}$ J s; $m_0 = 9.11 \times 10^{-31}$ kg.

Solution: The Compton shift at 90° to the direction of incidence is

$$\Delta\lambda = \frac{h}{m_0 c} = \frac{6.62 \times 10^{-34}}{9.11 \times 10^{-31} \times 3 \times 10^8}$$

$$= 2.422 \times 10^{-12} \text{ m} = 0.02422 \text{ Å}$$

(ii) The wavelength of the scattered radiation is

$$\lambda' = \lambda + \Delta\lambda = 1 + 0.02422 = 1.02422 \text{ Å}$$

Energy of recoiling electron = loss in energy of radiation $= \dfrac{hc}{\lambda} - \dfrac{hc}{\lambda'}$

$$= 6.62 \times 10^{-34} \times 3 \times 10^{8} \left[\frac{1}{10^{-10}} - \frac{1}{1.02422 \times 10^{-10}}\right]$$

$$= 4.696 \times 10^{-17}\ \text{J}$$

6.9 Relativistic Doppler effect

Let us consider two inertial reference frames S and S′, where S′ is moving with velocity v relative to S along the common x-axis.

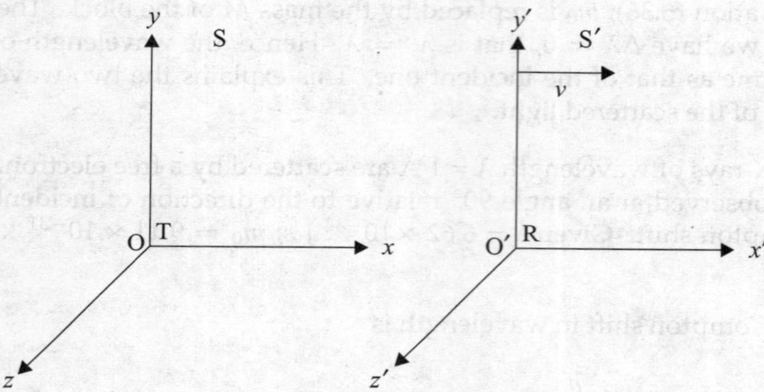

Suppose a transmitter T and a receiver R are situated at origins O and O′, respectively. Let two light signals or pulses be transmitted at time $t = 0$ and $t = T$, where T is the proper time period of light pulse, so that $dt = T$. Suppose $dt' = T'$ be the interval between the reception of these two pulses by the receiver in S′. From the Lorentz transformation

$$x = \frac{x' + vt'}{\sqrt{1-\beta^2}} \qquad \text{we have} \qquad dx = \frac{dx' + v\,dt'}{\sqrt{1-\beta^2}}$$

where $\beta = v/c$. Since the observer continues to be at O′ all the time, the distance travelled by him in S′ is zero ($dx' = 0$). Thus, we have

$$dx = \frac{v\,dt'}{\sqrt{1-\beta^2}} = \frac{vT'}{\sqrt{1-\beta^2}} \tag{6.37}$$

This equation shows that the second pulse has to travel a large distance by dx than that travelled by the first pulse in S along x-axis to be able to reach at O′ in S′. Also, from the Lorentz transformation

$$t = \frac{t' + vx'/c^2}{\sqrt{1-\beta^2}} \qquad \text{we have} \qquad dt = \frac{dt' + v\,dx'/c^2}{\sqrt{1-\beta^2}} = \frac{dt'}{\sqrt{1-\beta^2}}$$

Thus, we have

$$dt = \frac{T'}{\sqrt{1-\beta^2}} \tag{6.38}$$

This relation includes both the actual time period T of the pulses and the time taken (dx/c) by the second pulse to cover the extra distance dx in S. Now, we have

$$dt = T + \frac{dx}{c} \tag{6.39}$$

Using equations (6.37) and (6.38) in (6.39), we have

$$\frac{T'}{\sqrt{1-v^2/c^2}} = T + \frac{vT'}{c\sqrt{1-v^2/c^2}}$$

On solving this equation, we have

$$T = \frac{T'\left(1-v/c\right)}{\sqrt{1-v^2/c^2}} = T'\sqrt{\frac{1-v/c}{1+v/c}}$$

If ν and ν' are the actual (laboratory) and observed frequencies of light pulses, respectively, we have

$$T = \frac{1}{\nu} \qquad \text{and} \qquad T' = \frac{1}{\nu'}$$

and thus we have

$$\frac{1}{\nu} = \frac{1}{\nu'}\sqrt{\frac{1-v/c}{1+v/c}} \qquad \text{or} \qquad \nu' = \nu\sqrt{\frac{1-v/c}{1+v/c}} \tag{6.40}$$

For the actual (laboratory) and observed wavelengths λ and λ', we have relations

$$\nu = \frac{c}{\lambda} \qquad \text{and} \qquad \nu' = \frac{c}{\lambda'} \tag{6.41}$$

Using equation (6.41) in (6.40), we have

$$\frac{c}{\lambda'} = \frac{c}{\lambda}\sqrt{\frac{1-v/c}{1+v/c}} \qquad \text{or} \qquad \lambda' = \lambda\sqrt{\frac{1+v/c}{1-v/c}} \tag{6.42}$$

In equations (6.40) and (6.42), the velocity v is positive when the observer is receding away from the source and is negative when the observer is approaching the source. Thus, during recession, the apparent frequency ν' decreases and the apparent wavelength λ' increases. This is often referred to as the red shift. On the other side, during the approach, the apparent frequency ν' increases and the apparent wavelength λ' decreases. This is often referred to as the blue shift. This Doppler effect is the longitudinal Doppler effect, as the observations are made along the direction of motion of the light source.

6.9.1 Non-relativistic case

For smaller velocities ($v << c$), equation (6.40) gives

$$\nu' = \nu\left(1-\frac{v}{c}\right)^{1/2}\left(1+\frac{v}{c}\right)^{-1/2}$$

$$= \nu\left(1-\frac{v}{2c}\right)\left(1-\frac{v}{2c}\right) = \nu\left(1-\frac{v}{c}\right)$$

Here, square and higher-order terms are neglected. Thus, we have

$$\frac{\nu-\nu'}{\nu} = \frac{v}{c} \qquad \text{or} \qquad \frac{\Delta\nu}{\nu} = \frac{v}{c}$$

where $\Delta\nu = \nu - \nu'$. For smaller velocities ($v << c$), equation (6.42) gives

$$\lambda' = \lambda\left(1+\frac{v}{c}\right)^{1/2}\left(1-\frac{v}{c}\right)^{-1/2}$$

$$= \lambda\left(1+\frac{v}{2c}\right)\left(1+\frac{v}{2c}\right) = \lambda\left(1+\frac{v}{c}\right)$$

Here, square and higher-order terms are neglected. Thus, we have

$$\frac{\lambda'-\lambda}{\lambda} = \frac{v}{c} \qquad \text{or} \qquad \frac{\Delta\lambda}{\lambda} = \frac{v}{c}$$

where $\Delta\lambda = \lambda' - \lambda$. In equations (6.40) and (6.42), the velocity v is positive when the observer is receding away from the source and is negative when the observer is approaching the source. Thus, during recession, the apparent frequency ν' decreases and the apparent wavelength λ' increases. This is often referred to as the red shift. On the other side, during the approach, the apparent frequency ν' increases and the apparent wavelength λ' decreases. This is often referred to as the blue shift.

Exercise 6.29: Calculate wavelength shift for H_α (6563 Å) line emitted by a star receding away from the earth with a velocity $0.1c$.

Solution: The wavelength observed on the surface of the earth is

$$\lambda' = \lambda\sqrt{\frac{1+v/c}{1-v/c}} = 6563\sqrt{\frac{1+0.1c/c}{1-0.1c/c}} = 7256\ \text{Å}$$

6.10 Four-vectors

The coordinates of an event (ct, x, y, z) can be considered as the components of a four-dimensional radius vector (or four-radius vector) in a four-dimensional space. The components are denoted by x^i, where the index i can take the values 0, 1, 2, 3, so that

$$x^0 = ct \qquad x^1 = x \qquad x^2 = y \qquad x^3 = z$$

The Lorentz transformations are

$$x'^0 = \frac{x^0 - \beta x^1}{\sqrt{1-\beta^2}}; \qquad x'^1 = \frac{x^1 - \beta x^0}{\sqrt{1-\beta^2}}; \qquad x'^2 = x^2; \qquad x'^3 = x^3$$

where $\beta = v/c$. The inverse Lorentz transformations are

$$x^0 = \frac{x'^0 + \beta x'^1}{\sqrt{1-\beta^2}}; \qquad x^1 = \frac{x'^1 + \beta x'^0}{\sqrt{1-\beta^2}}; \qquad x^2 = x'^2; \qquad x^3 = x'^3$$

In general, A^i (with components A^0, A^1, A^2, A^3) is a four-dimensional vector (or four-vector) whose components follow the Lorentz transformations. That is, we have

$$A'^0 = \frac{A^0 - \beta A^1}{\sqrt{1-\beta^2}}; \qquad A'^1 = \frac{A^1 - \beta A^0}{\sqrt{1-\beta^2}}; \qquad A'^2 = A^2; \qquad A'^3 = A^3$$

and

$$A^0 = \frac{A'^0 + \beta A'^1}{\sqrt{1-\beta^2}}; \qquad A^1 = \frac{A'^1 + \beta A'^0}{\sqrt{1-\beta^2}}; \qquad A^2 = A'^2; \qquad A^3 = A'^3$$

The component A^0 is known as the time component, and A^1, A^2, A^3 are known as the space components of the four-vector.

6.10.1 Four-velocity

From the relation

$$ds^2 = c^2 dt^2 - dx^2 - dy^2 - dz^2$$

we get

$$ds = c\, dt \left(1 - \frac{v^2}{c^2}\right)^{1/2} = \gamma c\, dt$$

where we have used

$$v^2 = \frac{dx^2 + dy^2 + dz^2}{dt^2} \qquad \text{and} \qquad \gamma = \left(1 - \frac{v^2}{c^2}\right)^{1/2}$$

Four-dimensional velocity (four-velocity) u^i of a particle is

$$u^i = \frac{dx^i}{ds} \qquad \text{or} \qquad u^i = \frac{1}{c\gamma}\frac{dx^i}{dt}$$

Thus, we have

$$u^0 = \frac{1}{c\gamma}\frac{dx^0}{dt} = \frac{1}{\gamma} \qquad u^1 = \frac{1}{c\gamma}\frac{dx^1}{dt} = \frac{v_x}{c\gamma}$$

$$u^2 = \frac{1}{c\gamma}\frac{dx^2}{dt} = \frac{v_y}{c\gamma} \qquad u^3 = \frac{1}{c\gamma}\frac{dx^3}{dt} = \frac{v_z}{c\gamma}$$

Note that the four-velocity is a dimensionless quantity.

6.10.2 Four-momentum

Four-dimensional momentum (four-momentum) p^i of a particle is

$$p^i = m_0 c u^i$$

where m_0 is the rest mass of the particle and u^i the four-velocity. Using the expression for u^i, we have

$$p^0 = m_0 c u^0 = \frac{m_0 c}{(1 - v^2/c^2)^{1/2}} = \frac{m}{c} = \frac{E}{c}$$

where $E = mc^2$ is the energy of the particle. Now,

$$p^1 = m_0 c u^1 = \frac{m_0 v_x}{(1 - v^2/c^2)^{1/2}} = m v_x = p_x$$

$$p^2 = m_0 c u^2 = \frac{m_0 v_y}{(1 - v^2/c^2)^{1/2}} = m v_y = p_y$$

$$p^3 = m_0 c u^3 = \frac{m_0 v_z}{(1 - v^2/c^2)^{1/2}} = m v_z = p_z$$

6.10.3 Lorentz transformations for velocity

The Lorentz transformations are

$$u'^0 = \frac{u^0 - \beta u^1}{\sqrt{1-\beta^2}}; \qquad u'^1 = \frac{u^1 - \beta u^0}{\sqrt{1-\beta^2}}; \qquad u'^2 = u^2; \qquad u'^3 = u^3$$

where $\beta = v/c$. Hence, we have

$$\frac{v'_x}{c\gamma} = \frac{v_x/c\gamma - v/c\gamma}{\sqrt{1-\beta^2}}; \qquad v'_y = v_y; \qquad v'_z = v_z$$

or

$$v'_x = \frac{v_x - v}{\sqrt{1-\beta^2}}; \qquad v'_y = v_y; \qquad v'_z = v_z$$

The inverse Lorentz transformations are

$$u^0 = \frac{u'^0 + \beta u'^1}{\sqrt{1-\beta^2}}; \qquad u^1 = \frac{u'^1 + \beta u'^0}{\sqrt{1-\beta^2}}; \qquad u^2 = u'^2; \qquad u^3 = u'^3$$

Hence, we have

$$\frac{v_x}{c\gamma} = \frac{v'_x/c\gamma + v/c\gamma}{\sqrt{1-\beta^2}}; \qquad v_y = v'_y; \qquad v_z = v'_z$$

or

$$v_x = \frac{v'_x + v}{\sqrt{1-\beta^2}}; \qquad v_y = v'_y; \qquad v_z = v'_z$$

6.10.4 Lorentz transformations for momentum

The Lorentz transformations are

$$p'^0 = \frac{p^0 - \beta p^1}{\sqrt{1-\beta^2}}; \qquad p'^1 = \frac{p^1 - \beta p^0}{\sqrt{1-\beta^2}}; \qquad p'^2 = p^2; \qquad p'^3 = p^3$$

Hence, we have

$$\frac{E'}{c} = \frac{E/c - \beta p_x}{\sqrt{1-\beta^2}}; \qquad p'_x = \frac{p_x - \beta E/c}{\sqrt{1-\beta^2}}; \qquad p'_y = p_y; \qquad p'_z = p_z$$

or

$$E' = \frac{E - \beta c p_x}{\sqrt{1-\beta^2}}; \qquad p'_x = \frac{p_x - \beta E/c}{\sqrt{1-\beta^2}}; \qquad p'_y = p_y; \qquad p'_z = p_z$$

The inverse Lorentz transformations are

$$p^0 = \frac{p'^0 + \beta p'^1}{\sqrt{1-\beta^2}}; \qquad p^1 = \frac{p'^1 + \beta p'^0}{\sqrt{1-\beta^2}}; \qquad p^2 = p'^2; \qquad p^3 = p'^3$$

Hence, we have

$$\frac{E}{c} = \frac{E'/c + \beta p_x}{\sqrt{1-\beta^2}}; \qquad p_x = \frac{p'_x + \beta E/c}{\sqrt{1-\beta^2}}; \qquad p_y = p'_y; \qquad p'_z = p_z$$

or

$$E = \frac{E' + \beta c p_x}{\sqrt{1-\beta^2}}; \qquad p_x = \frac{p'_x + \beta E/c}{\sqrt{1-\beta^2}}; \qquad p_y = p'_y; \qquad p_z = p'_z$$

Multiple-choice questions

1. The aim of Michelson - Morley experiment was to

   (i) prove existence of ether. (ii) measure speed of light.

   (iii) measure speed of earth relative to the ether. (iv) test isotropy of the medium.

   Ans. (iii)

2. The negative result of Michelson - Morley experiment was that

   (i) it could not prove the existence of ether. (ii) it could not measure speed of light.

   (iii) it could not find the shifting of fringes. (iv) it could not prove electromagnetic nature of light.

   Ans. (iii)

3. The postulates of the special theory of relativity are applicable to

   (i) accelerated frames. (ii) rotating frames.

   (iii) inertial frames. (iv) non-inertial frames.

   Ans. (iii)

4. The special theory of relativity was proposed by

(i) Michelson and Morley. (ii) Einstein.

(iii) Lorentz. (iv) Galileo.

Ans. (ii)

5. For Lorentz transformations, which of the following statements is wrong?

(i) Time is not absolute. (ii) Time and space are coupled.

(iii) Velocity is invariant. (iv) None of above

Ans. (iii)

6. Each of the two photons A and B are moving with velocity c in the directions opposite to each other. The velocity of B relative to A is

(i) $2c$ (ii) 0

(iii) c (iv) $c/2$

Ans. (iii)

7. Which of the following is not a consequence of Lorentz transformations?

(i) Electromagnetic nature of light (ii) Length contraction

(iii) Coupling of space and time (iv) Time dilation

Ans. (i)

8. For a particle moving with relativistic velocity, the kinetic energy is

(i) $mc^2 - m_0c^2$ (ii) $mv^2/2$

(iii) m_0c^2 (iv) $mc^2/2$

Ans. (i)

9. Which of the following relations is not valid for special theory of relativity?

(i) $E^2 - c^2p^2 = m^2c^4$ (ii) $E = mc^2$

(iii) $m = m_0/\sqrt{1 - v^2/c^2}$ (iv) $E = mv^2/2$

Ans. (iv)

10. In which of the following phenomena, the relation of energy–mass equivalence is not used?

(i) Binding energy of a nucleus (ii) Compton effect

(iii) Photoelectric effect (iv) None of them

Ans. (iii)

11. When ν and λ are frequency and wavelength, respectively, measured in laboratory, which of the following relations is not an expression for relativistic Doppler effect?

(i) $\nu' = \nu\sqrt{\frac{1-v/c}{1+v/c}}$ (ii) $\lambda' = \lambda\sqrt{\frac{1+v/c}{1-v/c}}$

(iii) $\nu' = \nu\left(1 - \frac{v}{c}\right)$ (iv) None of them

Ans. (iii)

Exercises

1. What do you mean by relativity? Two bodies A and B are moving with velocities of 50 km/hr and 60 km/hr, respectively. What can be the minimum and maximum relative velocities?

2. Describe the Michelson - Morley experiment and discuss the importance of its result.

3. What are basic postulates of the special theory of relativity?

4. Consider two reference frames S and S′, where S′ is moving with linear velocity v relative to that of S. Show that if two events occur simultaneously in one reference frame, they do not occur simultaneously in other reference frame.

5. Derive Lorentz transformations for coordinates. Show that when $v << c$, the Lorentz transformations reduce to the Galilean transformations.

6. Derive Lorentz transformations for velocity components.

7. Photon A is moving with velocity c along x-axis and photon B is moving with velocity c in the opposite direction. Calculate the velocity of A relative to that of B. [Ans. c]

8. A body A is moving with velocity $0.5c$ along x-axis and body B is moving with velocity $0.5c$ in the opposite direction. Calculate the velocity of A relative to that of B. [Ans. $0.8c$]

9. Derive an expression showing the equivalence of energy and mass. What is its limitation?

10. Show that mass of a body varies with its velocity. Derive the following expression, where symbols have their usual meanings:

$$m = \frac{m_0}{\sqrt{1 - v^2/c^2}}$$

11. An electron with rest mass m_0 is moving with velocity $0.8c$, where c is the speed of light. Calculate the ratio of kinetic energy to rest energy of electron.

12. What is the relativistic Doppler effect? Derive an expression for it.

13. Suppose an inertial reference frame S′ is moving with velocity v relative to another inertial reference frame S along the direction of positive x-axis. If v_x, v_y, v_z are components of velocity in S and v'_x, v'_y, v'_z the components of velocity in S′, obtain the Lorentz transformations for the components of velocity.

14. Suppose an inertial reference frame S′ is moving with velocity v relative to another inertial reference frame S along the direction of positive x-axis. If E is total energy and p_x, p_y, p_z the components of momentum in S, and E' the total energy and p'_x, p'_y, p'_z the components of momentum in S′, obtain the Lorentz transformations for the energy and momentum.

15. Write notes on the following:

    (i) Relativity
    (ii) Lorentz transformations for coordinates
    (iii) Special theory of relativity
    (iv) Length contraction
    (v) Time dilation
    (vi) Variation of mass with velocity
    (vii) Mass–energy equivalence
    (viii) Relativistic Doppler effect
    (ix) Lorentz transformations for velocity
    (x) Lorentz transformations for momentum
    (xi) Compton effect

CHAPTER 7

NON-INERTIAL REFERENCE FRAMES

So far, in our discussion, we have assumed the existence of an inertial reference frame and all the experiments and theories have been developed for inertial reference frames. Reality seems to be otherwise. In fact, there is no inertial reference frame. Our laboratories are situated on the earth and any reference frame attached to the earth is not inertial reference frame, owing to its revolution around the sun and rotation about its own axis. The earth has translational motion which is not uniform and rotational motion. It is therefore doubly non-inertial. However, we cannot discard everything developed so far. We have already discussed with suitable reasoning that for practical purposes, the earth may be taken as inertial reference frame. Nevertheless, it is worth to discuss about the motions in non-inertial reference frames. An inertial reference frame moves with constant linear velocity. Thus, the deviation from non-inertial can be in two ways: (i) reference frame is moving with accelerated linear velocity and (ii) reference frame is moving with angular velocity. We shall handle the anomaly in two steps. First, we shall see how to make corrections when our reference frame is moving with linear acceleration with respect to an inertial reference frame. Next, we shall account for corrections when our reference frame is rotating with respect to an inertial reference frame.

7.1 Linearly accelerating reference frame

Here, we shall look into the motion of a particle as seen from a reference frame S′ which is accelerated with respect to an inertial reference frame S. Here, we shall not account for any rotational motion.

7.1.1 Transformation of acceleration

Let us consider an inertial reference frame S and a non-inertial reference frame S′. The reference frame S′ is moving with a constant acceleration α with respect to the reference frame S, as shown in Figure 7.1. For convenience, we consider that the motion of S′ is along the x-axis. Suppose P is a moving particle of mass m whose coordinates measured in S are (x, y, z) and measured in S′ are (x', y', z') at an instant t. Now, we want to find the equations of transformation of coordinates, velocity components and acceleration components.

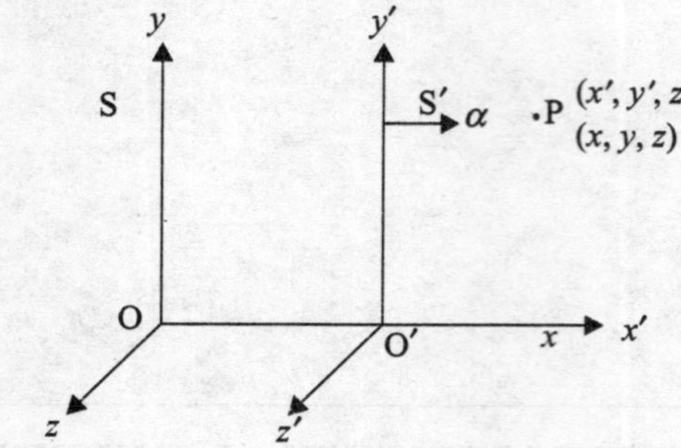

Figure 7.1: Non-inertial reference frame S′ is moving with constant acceleration α with respect to inertial reference frame S along x-axis.

Since the velocities are non-relativistic, we shall proceed in the same manner as the Galilean transformations were obtained. Suppose S′ is moving with constant acceleration $\alpha\hat{i}$. Let at $t = 0$, the two reference frames S and S′ were moving with the same velocity and the origins O and O′ of the reference frames S and S′, respectively, were coinciding. Thus, the relative velocity between S and S′ at $t = 0$ was zero. It obviously shows that the separation between O′ and O after a time t will be

$$\frac{1}{2}\alpha t^2$$

Thus, we have

$$x' = x - \frac{1}{2}\alpha t^2 \qquad y' = y \qquad z' = z \tag{7.1}$$

On differentiation of equation (7.1) with respect to t, we get

$$v'_x = v_x - \alpha t \qquad v'_y = v_y \qquad v'_z = v_z \tag{7.2}$$

where $v_x = dx/dt, v_y = dy/dt, v_z = dz/dt$ are components of velocity measured in S and $v'_x = dx'/dt$, $v'_y = dy'/dt$, $v'_z = dz'/dt$ are components of velocity measured in S′. On differentiation of equations (7.2) with respect to t, we get

$$a'_x = a_x - \alpha \qquad a'_y = a_y \qquad a'_z = a_z \tag{7.3}$$

where $a_x = dv_x/dt$, $a_y = dv_y/dt$, $a_z = dv_z/dt$ are components of acceleration measured in S and $a'_x = dv'_x/dt$, $a'_y = dv'_y/dt$, $a'_z = dv'_z/dt$ are components of acceleration measured in S′. Equations (7.2) and (7.3) can be expressed as

$$\vec{v}' = \vec{v} - \vec{\alpha}\, t \qquad \text{and} \qquad \vec{a}' = \vec{a} - \vec{\alpha} \tag{7.4}$$

Although the relations (7.4) have been derived for constant acceleration along x-axis, they are valid for constant acceleration along any direction.

7.1.2 Origin of pseudo force

Figure 7.2(a) shows trajectory APB for the motion of particle in inertial reference frame S. At a certain instance of time t, the particle has velocity $\vec{v}$ and acceleration $\vec{a}$ and it is subject to a force $\vec{F}$.

Figure 7.2(b) shows the motion of the same particle as seen from the accelerated reference frame S′. The velocity $\vec{v}'$ and acceleration $\vec{a}'$ of the particle seen in S′ are

$$\vec{v}' = \vec{v} - \vec{\alpha}\, t \quad \text{and} \quad \vec{a}' = \vec{a} - \vec{\alpha} \tag{7.5}$$

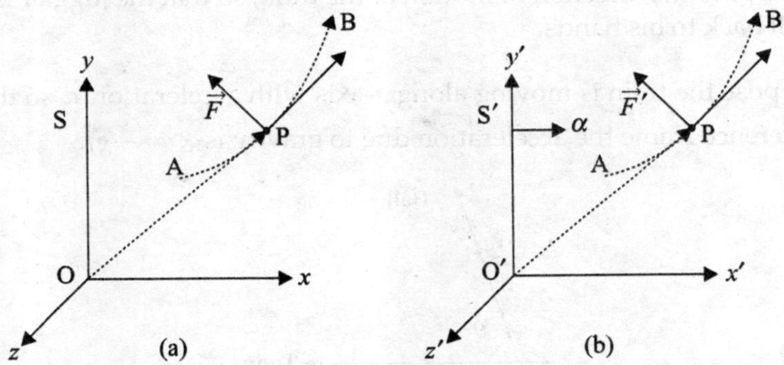

Figure 7.2: A particle P is moving along the path APB: (a) in inertial reference frame S and (b) in accelerated reference frame S′.

Newton's second law of motion in S is

$$\vec{F} = m\,\vec{a} \tag{7.6}$$

Using equation (7.5) in (7.6), we have

$$\vec{F} = m(\vec{a}' + \vec{\alpha})$$

Therefore, we have

$$m\,\vec{a}' = \vec{F} - m\,\vec{\alpha} \tag{7.7}$$

Equation (7.7) may be regarded as Newton's second law of motion in accelerated reference frame S′ when we are willing to replace $\vec{F} - m\,\vec{\alpha}$ by $\vec{F}'$, so that

$$m\,\vec{a}' = \vec{F}'$$

where

$$\vec{F}' = \vec{F} + \vec{F}_P$$

and the pseudo force $\vec{F}_P$ is

$$\vec{F}_P = -m\,\vec{\alpha}$$

Thus, we have added an extra force $\vec{F}_P$ to the true force $\vec{F}$, so that the same kind of equation of motion (Newton's second law of motion) is obtained in the accelerated reference frame S′. This extra force is known by different names: pseudo force, fictitious force (as it is purely imaginary without a real existence), inertial force, etc. It should be noted that an inertial force arises only in a non-inertial reference frame.

Exercise 7.1: Consider a superfast train moving with acceleration $\vec{\alpha}$ along the rail track. A juggler sitting on a chair in the train is making a show with a ball by throwing it into air and catching it back when it falls. During the period of acceleration of the train, he finds that the ball will not fall back to his hands unless it is thrown at some angle. Find the angle θ of throw relative to the direction of motion of the train, so that the juggler is successful in getting the ball back to his hands.

Solution: Suppose the train is moving along x-axis with acceleration $\vec{\alpha}$ so that $\vec{\alpha} = \alpha\hat{i}$. In the inertial reference frame the acceleration due to gravity is $\vec{g} = -g\hat{j}$.

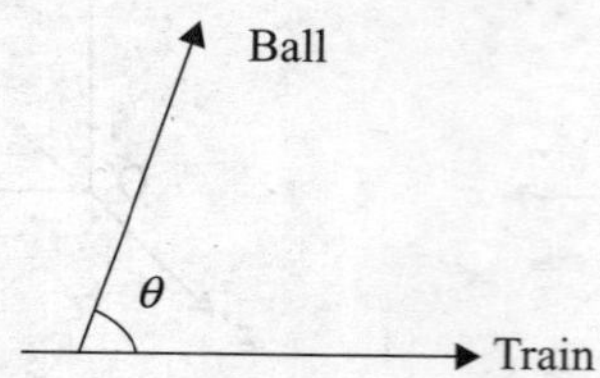

Thus, the acceleration due to gravity in the accelerated train is

$$\vec{g}' = \vec{g} - \vec{\alpha} = -g\hat{j} - \alpha\hat{i}$$

Suppose θ is the angle with x-axis along which the ball of mass m be thrown so that the juggler is successful in getting the ball back. Then, we have

$$\tan\theta = \frac{F_y}{F_x} = \frac{mg_y}{mg_x} = \frac{g_y}{g_x} = \frac{-g}{-\alpha} = \frac{g}{\alpha}$$

Exercise 7.2: Consider a superfast train moving with acceleration $\vec{\alpha}$ along the rail track. A pendulum bob is suspended from the ceiling of a coach. (i) Find the angle which the pendulum will make with the horizontal. (ii) Find the period of oscillation of pendulum.

Solution: Suppose the train is moving along x-axis with acceleration $\vec{\alpha}$ so that $\vec{\alpha} = \alpha\hat{i}$. In the inertial reference frame, the acceleration due to gravity is $\vec{g} = -g\hat{j}$.

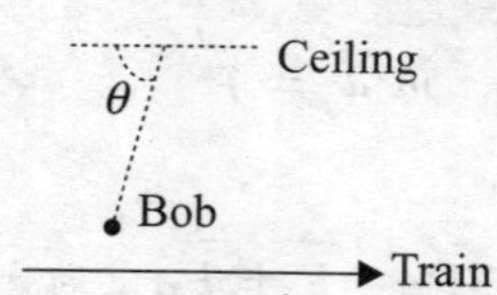

Thus, the acceleration due to gravity in the accelerated train is

$$\vec{g}' = \vec{g} - \vec{\alpha} = -g\hat{j} - \alpha\hat{i}$$

Suppose θ is the angle with x-axis along which the net acceleration due to gravity acts. Then, we have

$$\tan\theta = \frac{F_y}{F_x} = \frac{mg_y}{mg_x} = \frac{g_y}{g_x} = \frac{-g}{-\alpha} = \frac{g}{\alpha}$$

The magnitude of net acceleration due to gravity is

$$g' = \sqrt{g^2 + \alpha^2}$$

For a pendulum of length l, the time period T of oscillation of the pendulum is

$$T = 2\pi\sqrt{\frac{l}{g'}} = 2\pi\sqrt{\frac{l}{\sqrt{g^2+\alpha^2}}} = 2\pi\left(\frac{l^2}{g^2+\alpha^2}\right)^{1/4}$$

Exercise 7.3: Consider a resident of a multi-storied building riding an elevator for ascending and descending between ground floor (floor #0) and the top floor (floor #20). The ascend and descend is divided into six stages, first three stages for ascend and the last three stages for descend.

(1) Constant acceleration of 2 m/s^2 from floor #0 to #5.

(2) Constant velocity v from floor #5 to #14.

(3) Constant deceleration of α' from floor #14 to #20.

(4) Free fall from floor #20 to #16.

(5) Constant downward velocity u from floor #16 to #11.

(6) Constant deceleration of α'' from floor #11 to #0.

(i) Find out the values of v, u, α' and α''. (ii) If the resident is standing on a weighing machine during ascending and descending, what will be his weight as recorded on the machine in each of the six stages? Given height of each floor h = 3 m, mass of the resident m = 75 kg. For convenience, use the acceleration due to gravity in the inertial reference frame $g = 10$ m/s^2.

Solution: (i) For the stage (1), the distance travelled is $h = 5 \times 3 = 15$ m. Thus, the velocity v is expressed as

$$v^2 = 2\alpha h = 2 \times 2 \times 15 = 60$$

Thus, $v = \sqrt{60}$ = 7.746 m/s. For the stage (3), the distance travelled is $h = 6 \times 3 = 18$ m. Thus, we have

$$0 = v^2 - 2\alpha' h = 60 - 2\alpha' 18$$

Thus, $\alpha' = 1.667$ m/s^2. For the stage (4), the distance travelled is $h = 4 \times 3 = 12$ m. Thus, the velocity u is expressed as

$$u^2 = 2gh = 2 \times 10 \times 12 = 240$$

Thus, $u = \sqrt{240} = 15.492$ m/s. For the stage (6), the distance travelled is $h = 11 \times 3 = 33$ m. Thus, we have

$$0 = u^2 - 2\alpha''h = 240 - 2\alpha''33$$

Thus, $\alpha'' = 3.636\ \mathrm{m/s^2}$.

(ii) The earth's force of gravity in each stage is $mg = 75 \times 10 = 750$ N. It can also be expressed as 75 kg weight. The weight is[1]

$$w = m(g + \alpha)$$

Stage #1: $w = 750 + 75 \times 2 = 900$ N

Stage #2: $w = 750 + 0 = 750$ N

Stage #3: $w = 750 + 75 \times (-1.667) = 624.975$ N

Stage #4: $w = 750 + 75 \times (-10) = 0$ N

Stage #5: $w = 750 + 0 = 750$ N

Stage #6: $w = 750 + 75 \times (3.636) = 1022.7$ N

7.2 Rotating reference frame

Let us consider two reference frames S and S′ having a common origin O. Suppose S′ rotates about some axis with an angular velocity $\vec{\omega}$. It is a case of pure rotation.[2] Let the coordinates of a point P measured at some time t in S and S′ be (x, y, z) and (x', y', z'), respectively. Then, we have

$$\vec{r} = x\hat{i} + y\hat{j} + z\hat{k} \quad \text{and} \quad \vec{r}' = x'\hat{i}' + y'\hat{j}' + z'\hat{k}' \tag{7.8}$$

where $\hat{i}$, $\hat{j}$, $\hat{k}$ are the unit vectors along the axes in S and $\hat{i}'$, $\hat{j}'$, $\hat{k}'$ are the unit vectors along the axes in S′ (Figure 7.3). Since the point P is same, we have $\vec{r} = \vec{r}'$. Let us assume that S be a fixed reference frame and S′ rotates relative to S with a constant angular velocity $\vec{\omega}$.

---

[1] Let us first clarify the meaning of the term 'weight'. The rider is in equilibrium under the action of two forces: The effective gravity force $\vec{F}_g = m\,\vec{g}' = m(\vec{g} - \vec{\alpha})$ and the reaction force $\vec{R}$ from the top layer of the weighing machine (on which the rider is standing). Taking y-axis as pointing upwards, we have $\vec{g} = -g\hat{j}$, $\vec{\alpha} = \alpha\hat{j}$, so that $\vec{F}_g = -m(g + \alpha)\hat{j}$. We have

$$\vec{R} + \vec{F}_g = 0 \quad \text{or} \quad \vec{R} = -\vec{F}_g = m(g + \alpha)\hat{j}$$

It is the reaction force which we call the weight of the person. From the above, the reaction force is in the y-direction. We express it as $w\hat{j}$. Then

$$w = m(g + \alpha)$$

is the magnitude of the weight of the person, or simply the weight of the person as recorded by the weighing machine. It shows that $w > mg$ when α is positive (i.e. upwards) and $w < mg$ when α is negative (i.e. downwards).

[2] Here, we are not accounting for the translational motion. For a general motion consisting of translational plus rotational motions, we can use the principle of superposition.

Then the unit vectors $\hat{i}', \hat{j}', \hat{k}'$ in S′ are changing with time. However, their magnitudes are equal to 1 unit. The velocities $\vec{v}$ and $\vec{v}'$ in S and S′, respectively, are

$$\vec{v} = \dot{x}\,\hat{i} + \dot{y}\,\hat{j} + \dot{z}\,\hat{k} \tag{7.9}$$

and

$$\vec{v}' = \dot{x}'\,\hat{i}' + \dot{y}'\,\hat{j}' + \dot{z}'\,\hat{k}' \tag{7.10}$$

where a *dot* on a variable represents the differentiation with respect to time. Here, we know that the time t does not depend on the reference frame. That is, in the classical mechanics, the time is absolute ($t = t'$). Now, in S, we have

$$\vec{v} = \frac{\mathrm{d}\vec{r}}{\mathrm{d}t}$$

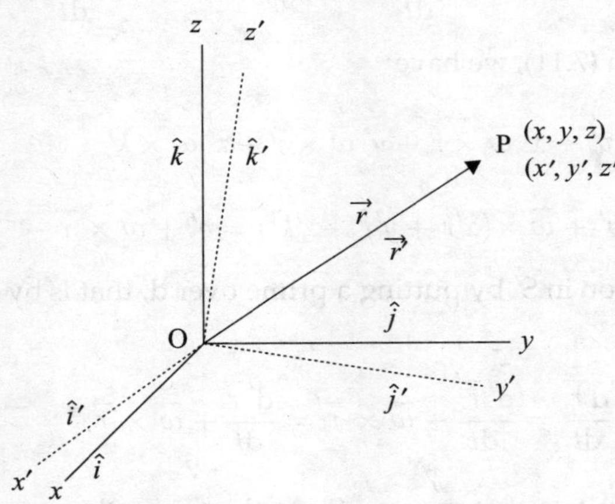

Figure 7.3: The reference frame S′ is rotating relative to S about some axis with a constant angular velocity $\vec{\omega}$.

At some instant t, coordinates of a point P in S and S′ are, respectively, (x, y, z) and (x', y', z'). Using $r = r'$, we have

$$\vec{v} = \frac{\mathrm{d}\vec{r}'}{\mathrm{d}t} = \frac{\mathrm{d}}{\mathrm{d}t}(x'\hat{i}' + y'\hat{j}' + z'\hat{k}')$$

$$= \dot{x}'\,\hat{i}' + \dot{y}'\,\hat{j}' + \dot{z}'\,\hat{k}' + x'\frac{\mathrm{d}\hat{i}'}{\mathrm{d}t} + y'\frac{\mathrm{d}\hat{j}'}{\mathrm{d}t} + z'\frac{\mathrm{d}\hat{k}'}{\mathrm{d}t}$$

$$= \vec{v}' + x'\frac{\mathrm{d}\hat{i}'}{\mathrm{d}t} + y'\frac{\mathrm{d}\hat{j}'}{\mathrm{d}t} + z'\frac{\mathrm{d}\hat{k}'}{\mathrm{d}t} \tag{7.11}$$

Here, we have used equation (7.10). Now, we would like to find out the values of $\mathrm{d}\hat{i}'/\mathrm{d}t$, $\mathrm{d}\hat{j}'/\mathrm{d}t$, $\mathrm{d}\hat{k}'/\mathrm{d}t$, in a fixed frame S in terms of the known quantities. It can be done by using the theorem (proved on page no. 219) that when a vector $\vec{A}$ of constant magnitude rotates with a constant angular velocity $\vec{\omega}$ in the anticlockwise direction around a fixed direction in a fixed frame, then we have

$$\frac{d\vec{A}}{dt} = \vec{\omega} \times \vec{A} \tag{7.12}$$

Since there is nothing special about the vector $\vec{A}$, equation (7.12) can be written in an operator form as

$$\frac{d()}{dt} = \vec{\omega} \times ()$$

where ( ) denotes any vector quantity having constant magnitude. Thus, we have

$$\frac{d\hat{i}'}{dt} = \vec{\omega} \times \hat{i}' \qquad \frac{d\hat{j}'}{dt} = \vec{\omega} \times \hat{j}' \qquad \frac{d\hat{k}'}{dt} = \vec{\omega} \times \hat{k}' \tag{7.13}$$

Using equation (7.13) in (7.11), we have

$$\vec{v} = \vec{v}' + x'\,\vec{\omega} \times \hat{i}' + y'\,\vec{\omega} \times \hat{j}' + z'\,\vec{\omega} \times \hat{k}'$$

$$= \vec{v}' + \vec{\omega} \times (x'\hat{i}' + y'\hat{j}' + z'\hat{k}') = \vec{v}' + \vec{\omega} \times \vec{r}' \tag{7.14}$$

Expressing differentiation in S′ by putting a prime over d, that is by writing d′/dt instead of d/dt for S′, we have

$$\frac{d\vec{r}}{dt} = \frac{d'\vec{r}'}{dt} + \vec{\omega} \times \vec{r}' = \frac{d'\vec{r}}{dt} + \vec{\omega} \times \vec{r} \tag{7.15}$$

where we have used the fact that $\vec{r}' = \vec{r}$. Since there is nothing special about a position vector $\vec{r}$, we get a general operator equation as

$$\frac{d(\,)}{dt} = \frac{d'(\,)}{dt} + \vec{\omega} \times (\,) \tag{7.16}$$

where ( ) is any vector quantity measured in S. For the velocity $\vec{v}$ measured in S, we have

$$\frac{d\vec{v}}{dt} = \frac{d'\vec{v}}{dt} + \vec{\omega} \times \vec{v} \tag{7.17}$$

Differentiating equation (7.14), we get

$$\frac{d'\vec{v}}{dt} = \frac{d'\vec{v}'}{dt} + \vec{\omega} \times \frac{d'\vec{r}'}{dt} = \vec{a}' + \vec{\omega} \times \vec{v}' \tag{7.18}$$

Using equation (7.18) in (7.17), we have

$$\frac{d\vec{v}}{dt} = \vec{a}' + \vec{\omega} \times \vec{v}' + \vec{\omega} \times \vec{v}$$

Thus, we have

$$\vec{a} = \vec{a}' + \vec{\omega} \times \vec{v}' + \vec{\omega} \times \vec{v} \tag{7.19}$$

Here, $\vec{a}$ and $\vec{a}'$ are acceleration of a body measured in S and S′, respectively. Using equation (7.14) in (7.19), we have

$$\begin{aligned}\vec{a} &= \vec{a}' + \vec{\omega} \times \vec{v}' + \vec{\omega} \times (\vec{v}' + \vec{\omega} \times \vec{r}') \\ &= \vec{a}' + 2\vec{\omega} \times \vec{v}' + \vec{\omega} \times (\vec{\omega} \times \vec{r}')\end{aligned} \tag{7.20}$$

Now, we would like to write Newton's second law of motion in a rotating frame S′. If $\vec{F}$ is an external force, then according to Newton's second law of motion in the fixed frame S, we have

$$\vec{F} = m\vec{a} \tag{7.21}$$

Using equation (7.20) in (7.21), we get

$$\vec{F} = m\vec{a}' + 2m(\vec{\omega} \times \vec{v}') + m\vec{\omega} \times (\vec{\omega} \times \vec{r}') \tag{7.22}$$

or

$$m\vec{a}' = \vec{F} - 2m(\vec{\omega} \times \vec{v}') - m\vec{\omega} \times (\vec{\omega} \times \vec{r}') \tag{7.23}$$

The external force does not depend on the reference frame. Since $m\vec{a}' \neq \vec{F}$, the rotating reference frame is not inertial one. Equation (7.23) gives the motion of a particle in a rotating frame S′. The term $-2m(\vec{\omega} \times \vec{v}')$ is known as the 'Coriolis force'. The force $-m\vec{\omega} \times (\vec{\omega} \times \vec{r}')$ is directed outwards and is known as the *centrifugal force*. The magnitude of the centrifugal force is $m\omega^2 r' (\equiv mv'^2/r')$. This is directed outwards of the curved path. This force is not a real force, but a fictitious one. Obviously, we can treat a rotating reference frame as inertial by adding a centrifugal force and a Coriolis force.

The Coriolis force depends on the velocity of the particle and is in the direction perpendicular to both $\vec{\omega}$ and $\vec{v}'$. Therefore, the Coriolis force does not do any work and it only changes the direction of motion of the moving body. If a body is stationary in the rotating frame, the Coriolis force is zero.

Theorem: If a vector $\vec{A}$ of constant magnitude rotates with a constant angular velocity $\vec{\omega}$ around a fixed direction in the anticlockwise direction in a fixed frame, then we have

$$\frac{d\vec{A}}{dt} = \vec{\omega} \times \vec{A}$$

Proof: Suppose a vector $\vec{A}$ of constant magnitude rotates in the anticlockwise direction, as shown in Figure 7.4, with a constant angular velocity $\vec{\omega}$ about an axis OM in the anticlockwise direction. Suppose at times t and $t + \delta t$, the vector is $\vec{A}(t)$ and $\vec{A}(t + \delta t)$, respectively.

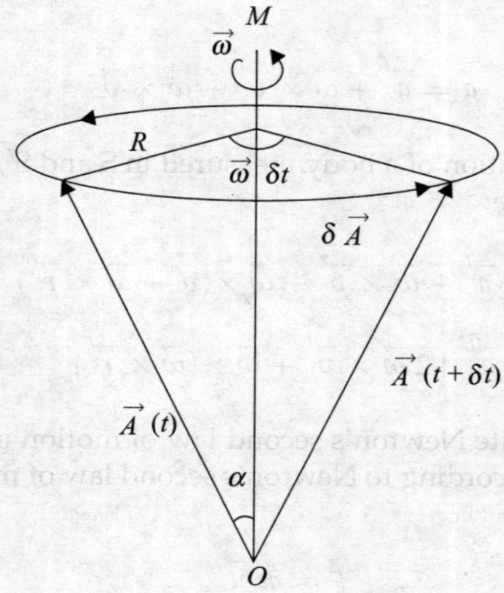

Figure 7.4: Vector $\overrightarrow{A}$ of constant magnitude is rotating about OM with a constant angular velocity $\overrightarrow{\omega}$ in the anticlockwise direction.

Although the vector changes from $\overrightarrow{A}(t)$ to $\overrightarrow{A}(t+\delta t)$, its magnitude does not change, that is $|\overrightarrow{A}(t)| = |\overrightarrow{A}(t+\delta t)|$. However, we have

$$\overrightarrow{A}(t+\delta t) = \overrightarrow{A}(t) + \delta\overrightarrow{A} \tag{7.24}$$

Since the vector is rotating with a constant angular velocity ω about OM, the angle traversed in time δt is $\omega\,\delta t$ and we have

$$|\delta\overrightarrow{A}| = R\,\omega\,\delta t$$

where $|\delta\overrightarrow{A}|$ denotes the magnitude of $\delta\overrightarrow{A}$. When $\delta t \to 0$, the vector $\delta\overrightarrow{A}$ is obviously along the tangent to the path. Form the figure, we have

$$\delta\overrightarrow{A} = (\overrightarrow{\omega} \times \overrightarrow{A})\,\delta t \tag{7.25}$$

Using equation (7.25) in (7.24), we have

$$\frac{\overrightarrow{A}(t+\delta t) - \overrightarrow{A}(t)}{\delta t} = \overrightarrow{\omega} \times \overrightarrow{A} \tag{7.26}$$

For the limit $\delta t \to 0$, we have

$$\frac{d\overrightarrow{A}}{dt} = \overrightarrow{\omega} \times \overrightarrow{A} \tag{7.27}$$

7.3 Spin motion of the earth

Although the shape of the earth is oblate, for practical purposes it is taken as spherical, as shown in Figure 7.5. Our earth spins about its own axis with time period of 24 hr. The points N and S of intersection of spin axis with the surface of the earth are known as the poles. Let us consider a plane passing through the centre of the earth and perpendicular to the spin axis. This plane intersects the earth's surface in the form of a circle, known as the equator. For a point P on the surface of the earth, we have angle λ, called the latitude. The latitude is 90° when the point is on a pole and 0° when the point is on the equator.

7.3.1 Centrifugal force on the earth

The centrifugal force on the earth depends on the latitude λ of the place. This is because the distance between the spin axis and the point P on the earth's surface varies with λ (Figure 7.5). The angular velocity of the earth is $\omega = 7.3 \times 10^{-5}$ rad/s. The centrifugal force has the maximum value at the equator, where radius of the earth is 6.4×10^6 m and the centrifugal acceleration at the equator is

$$\omega^2 r = (7.3 \times 10^{-5})^2 \times 6.4 \times 10^6 = 3.41 \times 10^{-2} \text{ m/s}^2$$

Hence, the maximum value of the centrifugal force on the surface of the earth is only about 3.5×10^{-3} times the acceleration due to gravity. The centrifugal force is zero on a pole of the earth. Hence, the centrifugal acceleration of a body on the earth is even smaller than the Coriolis acceleration and can be neglected in comparison to g.

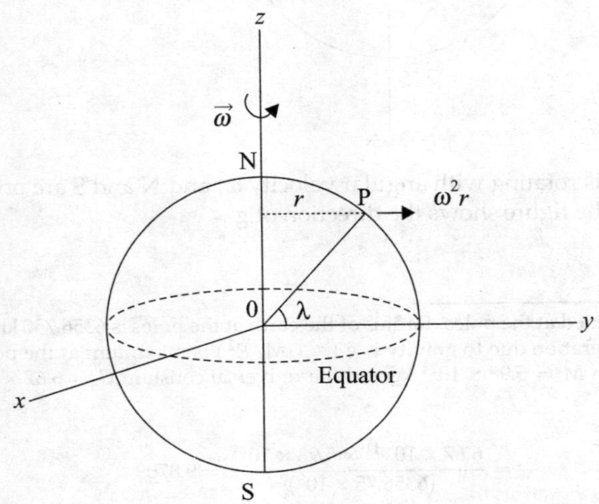

Figure 7.5: Centrifugal acceleration $\omega^2 r$ is directed away from the spin axis. λ is latitude of a place.

7.3.2 Effective force of gravity

Owing to the oblate shape of our earth, the acceleration due to gravity varies with latitude, being about 0.5% smaller at equator than at its poles.[3] Consider a point P on the earth's surface (Figure 7.6). If the earth was not rotating, acceleration due to gravity $\vec{g}$ at P would have acted along the direction PO (the line joining the point P to the centre O of the earth). Due to rotation of the earth, a centrifugal acceleration $(\vec{\omega} \times (\vec{\omega} \times \vec{r}))$ is superimposed on $\vec{g}$ and therefore an effective acceleration due to gravity $\vec{g}'$ at P is

$$\vec{g}' = \vec{g} - \vec{\omega} \times (\vec{\omega} \times \vec{r})$$

Here, $\vec{g}'$ does not pass through the centre O of the earth.

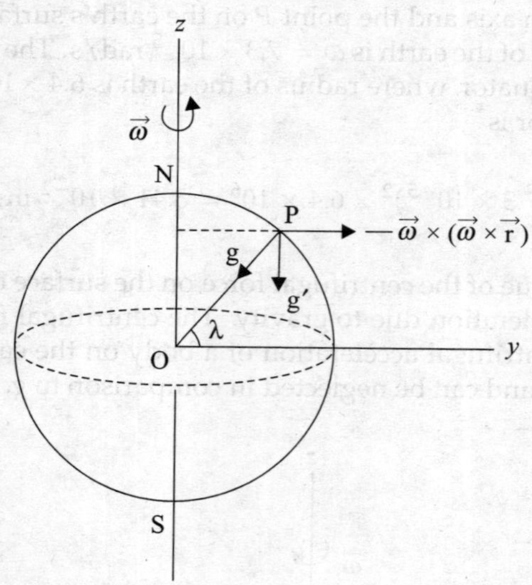

Figure 7.6: The earth is rotating with angular velocity ω, and N and S are north and south poles. λ is latitude of a place. The figure shows the direction of g′.

[3]Earth's sphere is flattened at the poles. Radius of the earth at the poles is 6356.750 km, whereas at the equator is 6378.135 km. The acceleration due to gravity is $g\ (\equiv GM/R^2)$ is maximum at the poles and minimum at the equator. Mass of the earth $M = 5.98 \times 10^{24}$ kg and the universal constant $G = 6.67 \times 10^{-11}$ N-m^2/kg^2. At the pole, we have

$$g = \frac{6.67 \times 10^{-11} \times 5.98 \times 10^{24}}{(6356.75 \times 10^3)^2} = 9.8709$$

At the equator, we have

$$g = \frac{6.67 \times 10^{-11} \times 5.98 \times 10^{24}}{(6378.135 \times 10^3)^2} = 9.8048$$

7.3.3 Deflection of a falling particle

Consider a particle of mass m is falling freely from a height h at a place having the latitude λ. We can estimate horizon deflection of the particle due to the Coriolis force.

Set the unit vectors $\hat{\imath}$, $\hat{\jmath}$, $\hat{k}$ as shown in Figure 7.7. Here, $\hat{\imath}$ and $\hat{\jmath}$ are in the plane of the surface of the earth and $\hat{k}$ perpendicular to the surface of the earth. Let P be a point on the earth's surface at latitude λ. Suppose a particle falls vertically from a height h above the earth's surface. As the particle does not have rotational motion, since it is falling freely, there is no centrifugal acceleration. After accounting for the Coriolis acceleration, the resultant acceleration $\vec{a}'$ is

$$\vec{a}' = \vec{g} - 2\,\vec{\omega} \times \vec{v}'$$

Components of $\vec{\omega}$ are $(-\omega\cos\lambda, 0, \omega\sin\lambda)$

Components of $\vec{g}$ are $(0, 0, -g)$

Actually, $\vec{a}'$ has components $(\ddot{x}', \ddot{y}', \ddot{z}')$ in the rotating frame attached to the earth. But, because the Coriolis force is very weak as compared to $\vec{g}$, we note that at the time t, the components of velocity are $\dot{x}' \approx 0$, $\dot{y}' \approx 0$, $\dot{z}' \approx -gt$. To a rather good approximation, the components of $\vec{v}'$ can be taken as $(0, 0, -gt)$. Thus, we have

$$\vec{\omega} \times \vec{v}' = \begin{vmatrix} \hat{\imath}' & \hat{\jmath}' & \hat{k}' \\ -\omega\ \cos\ \lambda & 0 & \omega\ \sin\ \lambda \\ 0 & 0 & -gt \end{vmatrix}$$

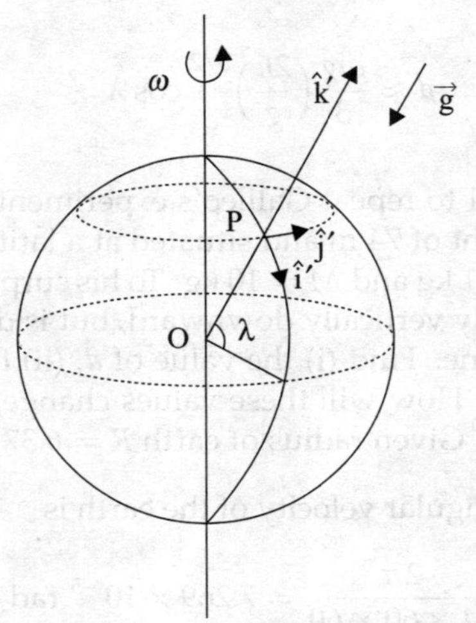

Figure 7.7: Unit vectors $\hat{\imath}$, $\hat{\jmath}$, $\hat{k}$ are set. λ is latitude of a place.

Thus, $\vec{\omega} \times \vec{v}' = -gt\omega\cos\lambda\,\hat{\jmath}'$ and therefore, we have

$$\vec{a}' = -g\,\hat{k}' + 2gt\omega\cos\lambda\,\hat{\jmath}'$$

Hence, the acceleration components are

$$\ddot{x}' = 0 \qquad \ddot{y}' = 2\omega g t \cos\lambda \qquad \ddot{z}' = -g$$

On integration, we get

$$\dot{x}' = c_1 \qquad \dot{y}' = \omega g t^2 \cos\lambda + c_2 \qquad \dot{z}' = -gt + c_3$$

where c_1, c_2, c_3 are constants of integration. Applying the condition that at $t = 0$, we have $\dot{x}' = 0$, $\dot{y}' = 0$, $\dot{z}' = 0$, we get $c_1 = 0$, $c_2 = 0$, $c_3 = 0$. Hence, we have

$$\dot{x}' = 0 \qquad \dot{y}' = \omega g t^2 \cos\lambda \qquad \dot{z}' = -gt$$

On integration, we get

$$x' = c_4 \qquad y' = \frac{1}{3}\,\omega g t^3 \cos\lambda + c_5 \qquad z' = -\frac{1}{2}\,g t^2 + c_6$$

where c_4, c_5, c_6 are constants of integration. Applying the condition that at $t = 0$, we have $x' = 0$, $y' = 0$, $z' = h$, we get $c_4 = 0$, $c_5 = 0$, $c_6 = h$. Using these values of the constants, we have

$$x' = 0 \qquad y' = \frac{\omega g t^3}{3}\cos\lambda \qquad z' = -\frac{1}{2}\,g t^2 + h$$

On putting $z' = 0$, we get the flight time $t = \sqrt{2h/g}$. Using $t = \sqrt{2h/g}$ in the expression for y', the northward deflection d in Northern Hemisphere, when particle falls freely from a height h at a latitude λ is

$$d \approx \frac{\omega g}{3}\left(\frac{2h}{g}\right)^{3/2} \cos\lambda$$

Exercise 7.4: A scientist tried to repeat Galileo's experiment by throwing a ball from the top of a Minaret having height of 74 m and situated at a latitude $\lambda = 29°\ 30'$. He dropped two iron balls of masses $m = 1$ kg and $M = 10$ kg. To his surprise, he finds that the first ball of mass m does not fall exactly vertically downward, but is deflected by a small amount d with respect to the vertical line. Find (i) the value of d, (ii) the magnitudes of centrifugal force and Coriolis force. (iii) How will these values change when he performs the same experiment with the mass M. Given radius of earth $R = 6.37 \times 10^6$ m, $g = 9.8\ \text{m/s}^2$.

Solution: (i) The rotational angular velocity of the earth is

$$\omega = \frac{2\pi}{24 \times 60 \times 60} = 7.269 \times 10^{-5}\ \text{rad/s}$$

The latitude of place $\lambda = 29°\ 30' = 29.5°$. We have $\cos\lambda = 0.870$. The deflection d is

$$d = \frac{\omega g}{3}\left(\frac{2h}{g}\right)^{3/2} \cos\lambda = \frac{7.269 \times 10^{-5} \times 9.8}{3}\left(\frac{2 \times 74}{9.8}\right)^{3/2} \times 0.870$$

$$= 1.21 \times 10^{-2}\ \text{m}$$

(ii) The centrifugal force F_{cf} is

$$F_{cf} = m\omega^2 R \cos\lambda = 1 \times (7.269 \times 10^{-5})^2 \times 6.37 \times 10^6 \times 0.870$$

$$= 2.928 \times 10^{-2}\ \text{N}$$

Under free fall, with initial velocity zero, the velocity is $9.8t$, where t is the time. The Coriolis force is

$$F_{cor} = 2m\omega v \cos\lambda = 2 \times 1 \times 7.269 \times 10^{-5} \times 9.8t \times 0.870$$

$$= 1.240 \times 10^{-3} t\ \text{N}$$

Time of fall

$$T = \sqrt{\frac{2h}{g}} = \sqrt{\frac{2 \times 74}{9.8}} = 3.886\ \text{s}$$

Thus, the maximum Coriolis force is

$$F_{cor}^{max} = 1.240 \times 10^{-3} T = 1.240 \times 10^{-3} \times 3.886 = 4.819 \times 10^{-3}\ \text{N}$$

(iii) When we take ball of mass $M = 10$ kg, the answer in part (i) will not change at all. The values of centrifugal force and Coriolis force will increase by a factor of $M/m = 10$.

7.3.4 Foucault pendulum

We know that a pendulum oscillates in a plane. French Physicist Foucault conceptualized that the Coriolis force would rotate the plane of oscillation of a pendulum. His pendulum, shown in Figure 7.8 is an ordinary pendulum with a large length l more than 10 m. Such a large length is necessary to make the effect significant, so that it is observable. Thus, the Foucault pendulum is a dramatic demonstration of rotation of the earth.

In order to understand how the plane of oscillation of the Foucault's pendulum rotates, let us write down the equation of motion in a rotating frame. When we assume that the earth is not rotating, then we have standard equation of motion[4]

$$\vec{a} + k_1^2\,\vec{r} = 0 \tag{7.28}$$

[4]Restoring force is

$$f = -mg\sin\theta = -mg\theta = -mg\,\frac{r}{l}$$

Thus, the acceleration is

$$\vec{a} = -\frac{g}{l}\,\vec{r}$$

The negative sign shows that the displacement and acceleration are in opposite directions.

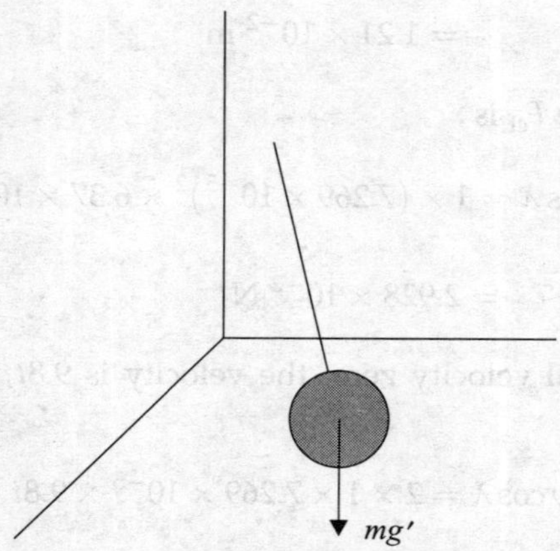

Figure 7.8: Foucault's pendulum.

where $k_1^2 = g/l$, and $\vec{a}$ and $\vec{r}$ are measured in the fixed reference frame (or in the inertial reference frame). Here, g is the acceleration due to gravity and l is the length of the pendulum.[5] Now, we account for the rotation of earth. That is, the pendulum is in the rotating frame. Now, the acceleration $\vec{a}'$ measured in S′ is related to the acceleration $\vec{a}$ measured in S as

$$\vec{a} = \vec{a}' + 2\,\vec{\omega} \times \vec{v}' \tag{7.29}$$

For the rotating frame, we have to use equation (7.29) in (7.28) and replace $\vec{g}$ by $\vec{g}'$. Thus, we have

$$\vec{a}' + 2\,\vec{\omega} \times \vec{v}' + k^2\,\vec{r}' = 0 \tag{7.30}$$

where $k^2 = g'/l$. Remember that $\vec{r} = \vec{r}'$. Since now all the quantities in equation (7.30) are measured in S′, in the further discussion, for convenience, let us drop prime from all quantities and from the unit vectors. Since the motion of the bob is almost linear, we have not accounted for the centrifugal acceleration. Let us consider a coordinate system where the z-axis is taken along the local vertical direction, that is opposite to the direction of g'. Let the point be P where the Foucault pendulum is kept at a latitude λ. Since the x and y motions of the bob are much larger than the z motion, we ignore the z motion. Thus,

---

[5]Equation (7.28) can be resolved into three components

$$\ddot{x} + k_1^2 x = 0 \qquad \ddot{y} + k_1^2 y = 0 \qquad \ddot{z} + k^2 z = 0$$

Since we are not interested in the motion along the z-axis, we ignore the last equation. By defining $\xi = x + iy$, the first two equations can be written as

$$\ddot{\xi} + k^2 \xi = 0$$

Solution of this equation is

$$\xi(t) = A\, e^{ik_1 t} + B\, e^{-ik_1 t}$$

Components of $\vec{\omega}$ are $(-\omega\cos\lambda, 0, \omega\sin\lambda)$ and components of $\vec{v}$ are $(\dot{x}, \dot{y}, 0)$

Hence,

$$\vec{\omega}\times\vec{v} = \begin{vmatrix} \hat{i} & \hat{j} & \hat{k} \\ -\omega\cos\lambda & 0 & \omega\sin\lambda \\ \dot{x} & \dot{y} & 0 \end{vmatrix}$$

$$= -\omega\sin\lambda\,\dot{y}\,\hat{i} + \omega\sin\lambda\,\dot{x}\,\hat{j} - \omega\cos\lambda\,\dot{y}\,\hat{k}$$

Resolving the equation of motion in the components, we have

$$\ddot{x} - 2\omega\,\dot{y}\sin\lambda + k^2 x = 0$$

$$\ddot{y} + 2\omega\,\dot{x}\sin\lambda + k^2 y = 0$$

$$\ddot{z} - 2\omega\,\dot{y}\cos\lambda + k^2 z = 0$$

Since we are not interested in the motion along the z-axis, we ignore the last equation. The first two equations are a set of coupled equations which can be solved by defining $\xi = x + iy$. The first two equations can be written as

$$\ddot{\xi} + 2i\omega\,\dot{\xi}\sin\lambda + k^2\xi = 0$$

Solution of this equation is[6]

$$\xi(t) = e^{-i\omega t\sin\lambda}\left[Ae^{it\sqrt{k^2+\omega^2\sin^2\lambda}} + Be^{-it\sqrt{k^2+\omega^2\sin^2\lambda}}\right]$$

$$= e^{-i2\pi t/\tau}\left[Ae^{it\sqrt{k^2+\omega^2\sin^2\lambda}} + Be^{-it\sqrt{k^2+\omega^2\sin^2\lambda}}\right]$$

But $k^2 >> \omega^2$. (There are many oscillations of the pendulum in 24 hr.) Then the part in square bracket represents usual oscillations of the pendulum. Prefactor represents effect of Coriolis force. Thus, the plane of oscillation of the Foucault pendulum rotates with the period

$$\tau = \frac{2\pi}{\omega\sin\lambda}$$

It shows that at the equator ($\lambda = 0$), the period of rotation is infinite, that is the plane of oscillation does not rotate. On the poles ($\lambda = \pm 90°$), the period of rotation is $\tau = 2\pi/\omega$, which is same as the period of rotation of the earth. At $\lambda = 30°$, the period of rotation is two times that of the earth. Hence, the Foucault pendulum is an experimental proof of the rotation of the earth relative to a fixed frame of reference.

[6]Using the operator D for differentiation, we have

$$D^2 + 2i\omega\sin\lambda D + k^2 = 0$$

Hence,

$$D = \frac{-2i\omega\sin\lambda \pm \sqrt{-4\omega^2\sin^2\lambda - 4k^2}}{2}$$

$$= -i\omega\sin\lambda \pm i\sqrt{k^2+\omega^2\sin^2\lambda}$$

Multiple-choice questions

1. Which of the following forces is fictitious?

   (i) Gravitational force (ii) Lorentz force

   (iii) Electric force (iv) Centrifugal force

   Ans. (iv)

2. Which of the following forces is fictitious?

   (i) Gravitational force (ii) Lorentz force

   (iii) Central force (iv) Coriolis force

   Ans. (iv)

3. A particle is falling freely from a distance h above the earth's surface. If spin velocity of earth is ω and latitude of the observer is λ, the deflection of particle on reaching the earth's surface is

   (i) $\frac{\omega g}{3}\left(\frac{2h}{g}\right)^{3/2} \cos\lambda$ (ii) $\frac{g}{3}\left(\frac{2h}{g}\right)^{3/2} \cos\lambda$

   (iii) $\frac{\omega g}{3}\left(\frac{2h}{g}\right)^{3/2}$ (iv) zero

   Ans. (i)

Exercises

1. Suppose a particle is moving in an inertial reference frame S and obeys Newton's second law of motion. What will happen when it is seen from another reference frame moving linearly with acceleration α with respect to S?
2. Explain how a pseudo force appears in case of an accelerated reference frame.
3. When we account for spin motion of earth, show that the gravity of earth does not pass through the centre of the earth.
4. Consider a superfast train moving with acceleration 9.8 m/s^2 along a rail track. A person sitting on a chair in the train is throwing a ball into air and catching it back when it falls. Find the angle θ of throw relative to the direction of motion of the train.
5. Consider a superfast train moving with acceleration 9.8 m/s^2 along a rail track. A simple pendulum of length $9.8\sqrt{2}$ is suspended from the ceiling of a coach. (i) Find the angle which the pendulum will make with the horizontal. (ii) Find the period of oscillation of pendulum.
6. Consider a resident of a multi-storied building riding an elevator for ascending and descending between ground floor (floor #0) and the top floor (floor #22). The ascend and descend is divided into six stages, first three stages for ascend and the last three stages for descend.

   (1) Constant acceleration of 2 m/s^2 from floor #0 to #6.

(2) Constant velocity v from floor #6 to #14.

(3) Constant deceleration of α' from floor #14 to #22.

(4) Free fall from floor #22 to #18.

(5) Constant downward velocity u from floor #18 to #11.

(6) Constant deceleration of α'' from floor #11 to #0.

(i) Find out the values of v, u, α' and α''. (ii) If the resident is standing on a weighing machine during ascending and descending, what will be his weight as recorded on the machine in each of the six stages? Given height of each floor $h = 4$ m, mass of the resident $m = 80$ kg. For convenience, use the acceleration due to gravity in the inertial reference frame as $g = 10\ \text{m/s}^2$.

7. A scientist performed an experiment by throwing a ball from the top of a Minaret having height of 80 m and situated at the latitude $\lambda = 32^\circ\ 40'$. He dropped an iron ball of mass $m = 2.5$ kg. He found that the ball does not fall exactly vertically downward, but is deflected by a small amount d with respect to the vertical line. Find (i) the value of d, (ii) the magnitudes of centrifugal force and Coriolis force. Given radius of earth $R = 6.37 \times 10^6$ m, $g = 9.8\ \text{m/s}^2$.

8. Write notes on the following:

   (i) Pseudo force
   (ii) Effective force of gravity
   (iii) Deflection of a falling particle
   (iv) Centrifugal force
   (v) Coriolis force

APPENDICES

The Greek alphabet

alpha	α		nu	ν	
beta	β		xi	ξ	Ξ
gamma	γ	Γ	omicron	$ø$	
delta	δ	Δ	pi	π	Π
epsilon	ϵ		rho	ρ	
zeta	ζ		sigma	σ	
eta	η		tau	τ	
theta	θ	Θ	upsilon	υ	
iota	ι		phi	ϕ	Φ
kappa	κ		chi	χ	
lambda	λ	Λ	psi	ψ	Ψ
mu	μ		omega	ω	Ω

Prefixes

Numerical values and abbreviation of some frequently used prefixes are as follows:

tera-	10^{12}	T	pico-	10^{-12}	p
giga-	10^{9}	G	nano-	10^{-9}	n
mega-	10^{6}	M	micro-	10^{-6}	μ
kilo-	10^{3}	k	milli-	10^{-3}	m

Linear motion versus angular motion

In case of angular motion, the direction of physical quantity is along the axis of rotation.

Linear motion		Angular motion	
Displacement	$\vec{r}$	Angle	θ
Velocity	$\vec{v}$	Angular velocity	$\vec{\omega}$
Acceleration	$\vec{a}$	Angular acceleration	$\vec{\alpha}$
Force	$\vec{F}$	Torque	$\vec{N}$
Mass	m	Moment of inertia	I
Momentum	$\vec{p}$	Angular momentum	$\vec{J}$

Table of values

1 radian	57.3° or 57° 18′
1 radian	3.44×10^3 minutes of arc
1 radian	2.06×10^5 seconds of arc
1 degree	1.75×10^{-2} rad
1 minute of arc	2.91×10^{-4} rad
1 seconds of arc	4.85×10^{-6} rad
1 angstrom	10^{-10} m
1 micron	10^{-6} m
1 light year	9.464×10^{15} m
1 parsec	3.08×10^{16} m
1 astronomical unit	1.40×10^{11} m
Radius of sun	6.96×10^8 m
Mass of sun	1.99×10^{30} kg
Mean radius of earth	6.37×10^6 m
Mass of earth	5.98×10^{24} kg
Mean radius of orbit of earth	1.49×10^{11} m
Period of revolution	1 year = 3.156×10^7 s
Period of rotation	24 hr = 8.64×10^4 s
Mass of moon	7.34×10^{22} kg
Radius of moon	1.74×10^6 m
Mean radius of orbit of moon	3.84×10^8 m
Period of revolution	2.36×10^6 s
Boltzmann constant k	1.38×10^{-23} J/K
Planck constant h	6.62×10^{-34} J-s
Stefan Boltzmann constant σ	5.67×10^{-8} W/m^2 k
speed of light in vacuum c	2.99725×10^8 m/s
gravitational constant G	6.67×10^{-11} N-m^2/kg^2
proton rest mass m_p	1.67×10^{-27} kg
electron rest mass m_e	9.1×10^{-31} kg
charge of proton e	1.6×10^{-19} C

Notations

1. $\sum_{i=1}^{n}$ denotes summation over what stands to the right of $\sum$ over all entries with range from $i = 1$ to $i = n$ with increment of 1.

2. $\sum_i$ denotes summation over what stands to the right of $\sum$ over all entries with all possible integer values of i with increment of 1.

3. $\sum_{ij}$ denotes double summation over all possible integer values of i and j with increment of 1 for both independently.

4. $\sum'_{ij}$ denotes double summation over all possible integer values of i and j except $i \neq j$ with increment of 1 for both independently.

Conics

1. Equation of ellipse in Cartesian coordinates is

$$\frac{x^2}{a^2} + \frac{y^2}{b^2} = 1$$

with eccentricity e such that $0 \leq e < 1$ and expressed as

$$b^2 = a^2(1 - e^2)$$

2. Circle is a special case of ellipse with $e = 0$ and $b = a$.

3. Equation of parabola in Cartesian coordinates is

$$y^2 = 4ax$$

with eccentricity $e = 1$.

4. Equation of hyperbola in Cartesian coordinates is

$$\frac{x^2}{a^2} - \frac{y^2}{b^2} = 1$$

with eccentricity $e > 1$ and expressed as

$$b^2 = a^2(e^2 - 1)$$

BIBLIOGRAPHY

Chandra S. 2011. *Classical Mechanics*. New Delhi: NAROSA Publishing House; Oxford (UK): Alpha Science International Ltd.

Chandra S. and M.K. Sharma. 2013. *An Introduction to Mathematical Physics*. New Delhi: NAROSA Publishing House; Oxford (UK): Alpha Science International Ltd.

Datta S. 2013. *Mechanics*. Chennai: Pearson.

Hans H.S. and S.P. Puri. 2003. *Mechanics, 2nd Edition*. New Delhi: Tata McGraw Hill Education.

Joshi A.W. and S. Chandra. 2010. *Solar System for Everyone*. New Delhi: National Council for Education Research & Training (NCERT).

Kakani S.L. and S. Kakani. 2008. *Engineering Physics*. New Delhi: CBS Publishers and Distributors.

Kittel C., W.D. Knight, M.A. Ruderman, C.A. Helmholz and B.J. Moyer. 2011. *Mechanics, 2nd Edition, 2nd Reprint*. New Delhi: Tata McGraw Hill Education.

Kleppner D. and R. Kolenkow. 2012. *An Introduction to Mechanics, 13th Reprint*. New Delhi: Tata McGraw Hill Education.

INDEX